ABINGDON COLLEGE LIBRARY

R04743W0805

D0120033

RIPPEL

26362

Transducers, Sensors, & Detectors

Transducers, Sensors, & Detectors

ROBERT G. SEIPPEL

RESTON PUBLISHING COMPANY, INC.
A Prentice-Hall Company
Reston, Virginia

ABINGDON COLLEGE OF FURTHER EDUCATION

Library

ABINGDON COLLEGE OF FURTHER EDUCATION
LIBRARY
26362
621.381548
S

Library of Congress Cataloging in Publication Data

Seippel, Robert G.
   Transducers, sensors & detectors.

    1. Transducers.   2. Detectors.   I. Title.
TK7881.2.S44 1983   621.3815'48   82–12197
ISBN 0–8359–7797–8

© 1983 by Reston Publishing Company, Inc.
A Prentice-Hall Company
Reston, Virginia 22090

All rights reserved. No part of this book may be reproduced in any way or by
any means, without permission in writing from the publisher.

5 7 9 10 8 6 4

Printed in the United States of America

Contents

v

Preface

Transducers, sensors, and detectors are an integral part of our personal and business lives. We, as citizens and world-builders, monitor and measure everything from sunlight to the shaking of a machine. Our modern way of life has become so complex that we must constantly check on the world's parameters to ensure that everything is working smoothly. Equipment is made so precisely that deviations from normal operation may destroy it or cause its product to be a failure. We monitor heartbeats, rainfall, the flow of water in steam plants, the temperature of a rotating member on a machine, and the count of cereal boxes as they are filled on an assembly line. In fact, there are not many things we as human beings do not monitor or measure.

The motivation to write this book emerged from other research and suggestions from friends in industry. The title is a broad description of perhaps the largest field in instrumentation and electronics.

Research for the book took about two years. Reference data came from the only source available, the manufacturers. Dozens of them provided the author with data such as study papers, application and engineering notes, and manufacturing processes. This descriptive literature was studied in detail and pertinent information was gleaned to make it meaningful to the reader.

During research there were several lessons that were learned. First, there are thousands of people who manufacture sensors. Second, there are thousands of sensor types. Third, a title covering all of them would fill a library.

With these lessons in mind, the author set out to put the book together. Manufacturers provide details of description, operation, and function of their own equipment and have done so for years. However, no one has attempted to tie it all together. This book attempts to do just that. It is a compilation of descriptive data, supplied to the author by state-of-the-art manufacturers. The level of the book is set to expose the greatest number of readers to an explanation of a large variety of transducers on a layperson's level of understanding.

To write a book of this type, much assistance is necessary. My wife Hazel Seippel deserves credit for her typing and all-around everything. Joanne Bly, my illustrator, and Patricia Rayner, my editor, are also deserving of accolades.

A special thanks is given to the contributors of data and illustrations within the book. I have attempted to include everyone who has made special contributions in this publication endeavor. I sincerely hope I have left no one out, for without the contributors' help this book would not be possible.

Robert G. Seippel, Ph.D.

Acknowledgments:

Automatic Timing and Controls Co., King of Prussia, Pennsylvania
    Mrs. Chris Perkins
BBN Instruments Corp., Cambridge, Massachusetts
    Albert J. Wells, Jr. & Carl Nicolino
Bell & Howell, CEC Div., Pasadena, California
    Robert W. Myers
Bowmar/TIC, Inc., Newbury Park, California
    Kenneth J. Cox
Computer Instruments Corp. (CIC), Hempstead, L.I., New York
    Don Wilen
Delevan Electronics, Inc., Scottsdale, Arizona
    John Wagoner & Gary Karlson
EDO Western Corp., Salt Lake City, Utah
    Gordon L. Snow
Electro Corp., Sarasota, Florida
    Emil J. Jasik
Electronic Flo-Meters (EFM), Inc., Dallas, Texas
    Henry N. Murphy
Envirotech-Sparling, El Monte, California
    William E. Bidgood
Hy-Cal Engineering, Santa Fe Springs, California
    Alice E. Bowman
Interface, Inc., Scottsdale, Arizona
    D. A. Carrington
Kulite Semiconductor Products, Inc., Ridgefield, New Jersey
    Joseph R. Mallon, Jr.

Metrix Instrument Co., Houston, Texas
Peter C. Sundt
Omega Engineering, Inc. (an Omega Group Co.), Stamford, Connecticut
Deborah Yamin & Rita De Rubeis
Paroscientific, Inc., Redmond, Washington
Jerome M. Paros
PCB Piezotronics, Inc., Buffalo, New York
Robert W. Lally
Rechner Electronic Industries, Inc., Niagara Falls, New York
A. Ramsay
Rockwell International, Pittsburgh, Pennsylvania
Michael Palbus
Setra Systems, Inc., Natick, Massachusetts
William Stern
Sundstrand Data Control, Inc., Redmond, Washington
Susan M. Moore
Tektronix, Inc., Beaverton, Oregon
Julie Fulton
Texas Electronics, Inc., Dallas, Texas
J. R. Tozer
The Singer Co., Kearfott Div., Little Falls, New Jersey
Dave Katz
Tylan Corp., Torrance, California
Allan B. Freeman
United Detector Technology, Culver City, California
Mike Wolpert
Vernitech, Inc., Deer Park, New York
Eric Partnoy
Westinghouse Electric Corp., Pittsburgh, Pennsylvania
Gary G. Forcey
Yellow Springs Instrument Co., Yellow Springs, Ohio
Philip H. Chitty

Transducers, Sensors, & Detectors

Foreword

Take a simple structure manifesting some physical law or effect; calibrate it; attach it to your test object; allow the two structures to interact naturally. You have just completed a measuring transaction using a transducer. Applying a little science, intuition, and experience to this task helps insure the validity of the measurement and builds confidence in the accuracy of the result.

Sensing transducers, or sensors, as they are often called, are the sensory components of measuring systems, which are part of a broad field of technology called *instrumentation*. The task of selecting and using instruments is generally referred to as *measurement engineering*. The sensing process is called *transduction*. While experience and intuition are still involved in successful measurements, instrumentation today is more of a science than an art, thanks to a number of dedicated scientists and educators.

Instrumentation plays a vital role in our modern technological world. In our expanding technology, more and more tasks of an experimental nature are being encountered as engineers and scientists, coping with extreme environments, strive for higher energy levels and faster, safer, quieter, and more efficient or reliable operation—automatically. The flight of the first space shuttle, viewed remotely by millions, is a prime example of an experimental project involving a multitude of instruments and computers. Complete space probes such as the *Mariner* and *Voyager* and including their launching rockets are in a sense the instruments.

Instrumentation also plays an important role in creating, building, and operating modern fuel-efficient cars. Downsizing the structures, improving the mileage, and reducing pollution involve much instrumented testing. These cars are now at least partially assembled by computer-directed industrial robots, involving instrumented feedback control systems. When the cars are operating,

1

onboard sensors and computers automatically adjust controls for optimum performance and fuel economy. To keep these cars supplied with fuel, instrumentation-based exploration and refining processes thus involve many sensors.

Measurement engineering is also helping to improve the quality of life. Sensors are intimately involved in reducing pollution and improving health. In much the same way that doctors use an electrocardiograph to check your heart, industrial engineers now employ instruments to monitor the health of machines. They also test the behavior of machines with instrumented hammers similar to the way doctors test your reflexes. Completing the loop, doctors now use industrial-related instruments to test human phenomena. It is not uncommon today to hear a doctor say that when you're forced or pressured to do something, you experience stress and the strain shows. A whole new field of medicine, biofeedback, is developing based on sensing physiological stress and strain. This rapidly growing field of medical electronics mostly consists of instrumentation.

STRUCTURAL SENSORS

Since people in almost all facets of science and technology are involved in testing, there exist many different ways of viewing sensors. To the physiologist, sensor systems are modeled after human nature. When your fingers touch or move, built-in force, motion, and temperature sensors send impulses along the nerve fibers to the brain, which gives you a mental picture of what's happening. Transducers relate to the human senses, cables to the nerve fibers, and the oscilloscope or analyzer to the brain.

To the structural test engineer, popular stress and strain types of transducers are mechanical structures. In his mind, all of the technology and all of terminology of structural analysis applies to the transducer structure as well as to that of the test object. This view, if universally adapted, could eliminate many of the barriers to communications between experimentally and analytically oriented people. People would no longer view transducers as mystical devices.

Since transduction is intimately related to the nature of things, the field of instrumentation involves many physical laws and effects and incompasses almost all fields of technology. Structural technology teaches us that it is the nature of things, including sensors, to deflect, vibrate, resonate, interact, conduct sound, experience stress and strain, and transfer force and motion. Moreover, structures including sensors behave differently at low, medium, and high frequencies, manifesting static and dynamic modes of behavior. All structures have an infinite number of resonant frequencies. Structures resonate at

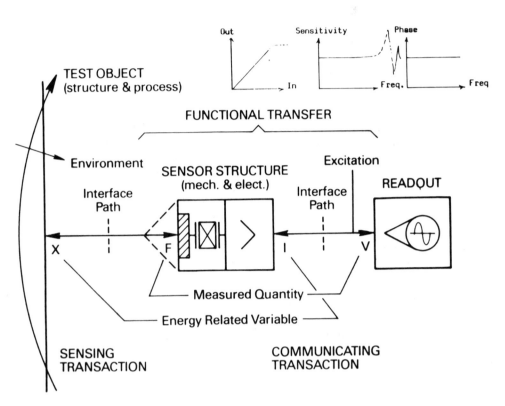

natural frequencies because elastic deflection moves mass and because inertia and elasticity impede motion in different ways, causing energy to alternate between kinetic and potential states. Finally, structural technology teaches that sensor structures can be mechanical, electrical, magnetic, fluidic, optical, chemical, thermal, and atomic, or a combination thereof.

Since test results and models created from test data reflect the combined behavior of the test object, fixturing, instrumentation, and computer structures, it is important to know something about the function, structure, and behavior of sensors. Popular stress and strain-gage-type transducers illustrate the principles involved.

TRANSDUCTION

As mentioned, it is the nature of structures to interact. Some time ago, a bright engineer or scientist observed that we can take the natural way in which structures interact and use it to sense and communicate information. Interaction between structures occurs as energy-transferring transactions, always involving two variables, effort and motion, whose product is energy. Thus, transduction employs an energy transfer process to sense and communicate information. But, because of the very nature of this transduction process, we can never measure a quantity without changing it.

The basic principle of transduction is a popular one today: *conserve energy*. Since transducers extract energy from the test object, the transduction process changes not only the quantity being measured, but the structure and process being tested as well. For valid measurements involving an insignificant amount of energy transfer, the rules are simple. When you are measuring an effort or potential variable like pressure, force, or voltage, select a transducer that greatly impedes motion or flow. When you are measuring motion or flow, choose a transducer that readily allows motion. In the electrical field, these rules are generally well understood. When measuring a voltage, you automatically choose a meter or oscilloscope with a specified input resistance very much higher than the output impedance of the device being tested.

The purpose of the measurement—research, test, control, or calibration—imposes different tolerances on this interaction effect called validity. The validity of a measurement can sometimes be checked experimentally by changing the sensor structure and observing the effect on the results.

SENSORS

A stress- or strain-gage sensor is an elastic structure that functions to transfer displacement caused by force into an electrical signal, more convenient for recording or processing. As discussed previously, the sensor structure is coupled in some way to the test object and the two structures are allowed to naturally interact. For force and motion measurement, the sensor structure can be a simple mechanical spring or a spring-mass system, called a stress- or strain-gage sensor. In these sensors, an integral or separate elastic sensing element transfers the deflection of the spring into an electrical signal. The sensing element can be made of resistive, piezoresistive, piezoelectric, capacitive, or inductive materials.

Behavior objectives for sensors are straight lines relating input and output and their ratio (sensitivity) with frequency. In other words, ideal sensors treat frequencies the same, amplitudes proportionally, do not delay the signal and, as discussed previously, do not appreciably change the quantity being measured. To achieve these objectives, sensors are usually operated well below their first resonant frequency, where the frequency response is reasonably flat. The accuracy of a sensor is usually specified as the deviation from some reference straight line. However, as long as the behavior is repeatable, modern computers and analyzers can correct for deviations by applying a separate calibration factor at each point in the amplitude and frequency range of interest, making the data appear as if they were measured with an ideal sensor.

The internal spring-mass, seismic structures of force, displacement, pressure, and acceleration sensors are essentially the same. Pressure transducers have an added diaphragm on bellows to transfer pressure into force. Accelerometers generally contain an enlarged mass.

Accelerometers implement Newton's laws of motion. They measure the force to automatically give their seismic mass the same motion as the test object to which they are attached. In piezoelectric accelerometers, the crystal elements perform a dual function. They act as a precision spring and generate an electrical signal proportional to their displacement. Servo accelerometers employ an internal feedback control circuit—displacement sensor, amplifier, and force generator—which behaves like a mechanical spring. Vibratory velocity sensors operate above their first resonant frequency and below any structural resonances where the mass stands still and the case moves about it. In velocity sensors, an inductive sensing element (a conductor in a magnetic field) functions to transfer the relative displacement into an electrical signal proportional to the velocity of the vibratory input motion. Accelerometers with built-in integrators do essentially the same thing.

CLASSIFICATION

Sensors are generally classified according to the variable being measured, the type of structure, the type of sensing element, and the behavior of the structure. For example, an instrument might be labeled a quartz piezoelectric, stress-gage, force transducer for measuring dynamic forces relative to an initial or average level. The terms "relative" and "dynamic" are generally omitted since they are characteristic of all electrostatic-type piezoelectric sensors.

The classification of a transducer as a stress- or strain-gage type depends upon which of the two parameters, stress or strain, the sensing element experiences the most during the measuring transaction, since both are force-actuated, defined the same, and experience stress and strain. This classification depends upon whether the elasticity of the sensing element or of the rest of the structure dominates the low-frequency behavior of the instrument. It is somewhat arbitrary, since either type of sensing element can measure force or displacement, depending upon how it is used.

ENVIRONMENT

Sensors, like human beings, are sensitive to all environmental inputs: pressure, force, motion, strain, sound, temperature, radiation, electric fields, magnetic fields, etc. They are generally designed to be quite sensitive to one environmental input, the measurand, and to be relatively insensitive to all others. To achieve the latter, various means of insulating, isolating, and compensating are often employed.

CALIBRATION

Calibration of a sensor involves applying a known input and measuring the output signal to determine the sensitivity of the instrument over specified amplitude and frequency ranges. Calibration of a transducer often involves its comparison with a reference transducer, whose calibration is traceable in some way to the National Bureau of Standards. In structural terminology, calibration is testing the functional transfer behavior of a sensor structure in controlled transactions and environments. The task of calibrating a sensor and testing the transfer function of a structure are essentially the same.

Sensors are often used under different conditions than employed during calibration, which may invalidate the measurement to some extent. In carefully controlled calibrating transactions, the measured quantity is usually associated with a very large energy source relative to the energy extracted for measurement. If, during use, the energy of the source is less, more interaction can occur, causing the quantity being measured to be changed appreciably. The solution to this dilemma is to choose an instrument that does not appreciably interact with your test object, or to program a computer to correct for the interaction.

ACCURACY

The very nature of the transduction process involving physical structures tends to distort, delay, and degrade the signal being transferred. Distortion of the signal is caused by nonlinearities, hysteresis, resonances, and environmental effects in the sensor structure. Any inadvertent energy storage in the process causes time delays between the output signal and the event. Time delays are mathematically modeled as phase shifts, which represents the time interval as a portion of the period of a complete 360-degree cycle at a given frequency. The normal sensing and processing of signals also always adds noise, which invariably degrades the information being transferred. Typical is electronic noise added during amplification. Environmental inputs and external power sources also add spurious noise. Averaging a number of signal segments in the time or frequency domain reduces the masking effect of random noise and improves the signal-to-noise ratio.

SYSTEMS

Sensor systems are generally composed of a number of elements packaged into component blocks. Most sensor systems contain a sensor, a conditioner, and an indicator, recorder, or analyzer. Each of these component blocks is a sensing, modifying, or output transducer. The transactional nature of the interaction between component blocks is sometimes depicted in diagrams with double-ended arrows. The conditioner may function to amplify, attenuate, standardize, integrate, filter, bias, or limit the signal and to regulate the power.

Between components, interface structures and paths, such as cables, modify the behavior of the system. Also, sensors seldom experience directly the quantity being measured. Typically, behavior modifying interface structures such as mounting pads or connecting acoustic passages separate the sensor and the test object.

TRENDS

Today, the extensive use of microelectronics and computers is having a profound effect on transducer designs. Three significant trends are apparent in advanced transducer products.

First, transducers are becoming much more sophisticated. Many of the signal-processing functions formerly relegated to external boxes are now being built inside the sensors as microelectronic modules. Such functions include amplifying, filtering, digitizing, analyzing, and storing, which make the sensor more directly compatible with computers and easier to use in quantity.

Second, digital-type sensors are becoming more popular. Sensors involving vibrating crystals or tuning forks that provide a varying frequency type of output signal transmit information more precisely and are more compatible with digital computers and recorders.

Third, fully integrated mechanical, electrical, electronic, and fluidic structures promise to significantly reduce the size and cost of sensors. Recently, mass-produced sensors for automotive applications have taken a big step in this direction.

The trend in systems is toward computer-controlled, multichannel instruments with convenience features such as auto-ranging, auto-calibration, overload monitoring, and auto-identification, to reduce the bookkeeping task. Such systems could transfer a multitude of simultaneous measurements directly into the computer, making possible near real-time behavior modeling of your test object.

SENSING NATURALLY

Almost from the time human beings were endowed with reflective and spiritual powers, they have fashioned instruments to help them cope with, adapt to, and explore their environment as well as their own nature, developing their human potential in the process. In the recent past, human beings have modeled machines after muscles, transducers after senses, and computers after the mind. Today, combining all of these mechanisms into sophisticated systems, we automatically control machines and processes, hominizing some of our own complex feedback behavior patterns (human nature). When applied to human processes, this technology is called *biofeedback*.

Viewing science and technology as a vital and natural part of the growth process of life represents both reality and humankind's current collective knowledge. This view gives an aesthetic value to a technology that reflects the order and grandeur of nature. In this light, a machine or an instrument, like a painting or composition, can be a beautiful creation, pleasing to the human mind and spirit.

When you consider that sensors are modeled after human nature, that they employ the natural way things interact to sense and communicate, that they often contain natural elements such as crystalline quartz, that they reflect the order and grandeur of nature, and that they have played a vital role in our evolutionary development, you can even feel good about making a single, solitary measurement.

R. W. LALLY
PCB PIEZOTRONICS, INC.
BUFFALO, NEW YORK.

1

An Introduction to Transducers, Sensors, and Detectors

GENERAL

A *transducer* is a device that converts one form of energy or physical quantity to another. Often this energy or physical quantity is "in form" or the same. The energy or stimulus determines the quantity of the signal.

A *sensor* is a device used to detect, measure, or record physical phenomena such as heat, radiation, and the like and to respond by transmitting the information, initiating changes, or operating controls.

A *detector* is a device used to sense the presence of something such as heat, radiation, or other physical phenomena.

As you can readily see, any difference between these three devices is an extremely thin line. There are other devices such as the gage and the pickup that essentially have the same definitions. For the purposes of this book we shall attempt to understand the transducer as the basic signal-gathering and transmitting device. However, we may transgress at times.

By developing our comprehension of the functions of a transducer, we will provide a broader definition of the device.

Measurement Functions

The functions of a transducer or sensor are deeply involved in measurement or control. In figure 1–1A, a sensor is attached directly to an indicator. This is the most basic of measurement systems. The sensor is acting as a sensing unit and also as a driver for an indicator pointer.

In figure 1–1B, a sensor responds to some physical quantity such as heat. Wires called extension wires, made from the same material as the sensor, extend

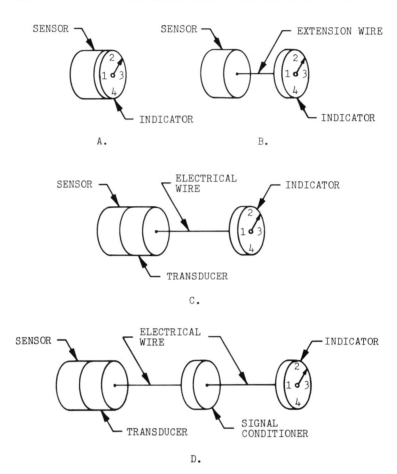

Figure 1-1   Measurement Functions

to an indicator. The indicator pointer responds to a change in the physical quantity felt at the sensor.

In another simple system shown in figure 1-1C, a transducer is used. The sensor responds to some physical quantity. The response is coupled to a compatible transducer which converts the sensor signal to an electrical signal. The electrical signal is transmitted by wire to an indicator pointer. The pointer responds to a change in the physical quantity felt at the sensor.

In a more complex system of measurement such as in figure 1-1D, a signal conditioner is used. The sensor responds to some physical quantity. The response is coupled to a compatible transducer which converts the sensor signal to an electrical signal. The electrical signal is transmitted to a signal conditioner whose job is to modify the signal for display. The modified signal is transmitted to the display for readout. The readout responds to a change in the physical quantity felt at the sensor.

This illustration provides a basis of measurement system organization.

Control Functions

The control function involves a system in which the transducer is an integral part. In figure 1-2, a sensor/transducer, signal conditioner, amplifier, and servo are in system to perform some operation. Figure 1-2 illustrates a fundamental control system. As with a measurement function, the sensor responds to some physical quantity. The response is coupled to a compatible transducer which converts the sensor signal to an electrical signal. The electrical signal is transmitted to a signal conditioner whose job is to modify the signal to drive a servo motor. The modified signal is amplified and directed to the motor of a servo. The servo then performs functions such as driving a control surface, opening a water passage, raising a platform, and similar operations.

Signal Analysis

The purpose of the transducer is to detect some signal phenomena. There are three major areas of signal phenomena. The first of these is *radiation;* here the transducer must be capable of detecting emission of radiation or reacting to radiation or the effects of it. The second major area of signal phenomena is *electrical.* This area involves detection of an electrical signal such as current or voltage. The third area is *physical* signal phenomena such as of mass or volume. In table 1-1, a list of signals is provided along with the science that deals with

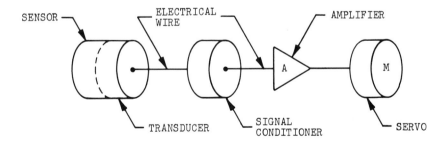

NOTE: SENSOR AND TRANSDUCER CAN BE THE SAME DEVICE.

Figure 1-2 Control Functions

TABLE 1-1. Signal Phenomena and Their Study Sciences

Signal Phenomena	Science Study
Radiation	
Emission	Spectroscopy
Scattering	Turbidimetry
Refraction	Refractometry
Absorption	Spectophotometry
Electrical	
Potential	Potentiometry
Current	Polarography
Resistance	Conductimetry
Physical	
Mass	Gravimetric Analysis
Volume	Volumetric Analysis
Rate	Kinetic Methods
Temperature	Thermal Conductivity

the method of measurement of each. This table could be expanded to include all the variations of light wavelengths, along with numerous electrical parameters and new study sciences in physical phenomena. Signal results are either qualitative or quantitative. Qualitative results are those whose molecular, atomic, physical, or functional features are found to be different from some standard. Quantitative results are measurable in numerical quantities.

Transducer Output Functions

You must remember that outputs from transducers may take the form of an electrical voltage or current or be nonelectrical in nature. Readout devices include such instruments as indicators, tape recorders, digital displays, or oscilloscopes. The signal, in the event the output signal is electrical, must be compatible with the readout device. The signal may be subjected to amplication, integration, or differentiation. It may also be added, subtracted, multiplied, divided, or in some other way modified.

Transducer Types

One of the functions of a transducer is to detect and convert a measurand into an electrical quantity. The measurand is the material or energy being measured, such as the amount of coal in a hopper. It is also worth noting that the measurand need not be energy, but could be material quantities.

There are basically two wide areas into which transducers are categorized. These are *active* and *passive*. Active transducers are those that generate a voltage or current as a result of some form of energy or force change. For instance, a thermocouple, when heated, generates an electrical signal that is proportional to the amount of heat applied. The passive transducer changes its properties when exposed to energy. For instance, photoresistors change resistance when exposed to light energy. Therefore the current through the photoresistor changes. In this event, we can again say that the amount of electrical output change is dependent on the amount of energy (light) to which the photoresistor was exposed.

There are other special transducer elements that may embody the active or passive elements for some combination of special interest. The electrokinetic transducer, for instance, operates with polar fluid maintained between two diaphragms to change the difference of potential between faces of its electrical interface. This element does not quite fit into the categories of active or passive transducers.

ACTIVE TRANSDUCER ELEMENTS

Active transducer elements are usually graded as those elements which generate a voltage as a result of some energy or force change.

The generation of signals is accomplished by six major methods. The methods include *electrostatics* and *chemical*. These two are not in general use as transducer element types. The other four methods are *electromechanical, photoelectrical, piezoelectrical,* and *thermoelectrical*. These are covered in some depth in the next several paragraphs.

Electromechanical Element

Relative motion through a magnetic field will produce a voltage at the ends of the conductor. That is, if a conductor passes through a field or the field is moved across a conductor, the motion will produce a voltage at the ends of the conductor. Such is still the case if both the conductor and the magnetic field are moving.

All these statements are representative of *Faraday's law of induction*. In the 1880s Michael Faraday, an English scientist, developed a simple machine in which a conductive disc was rotated in a magnetic field. Sliding contacts (brushes) were used to pick off the voltage from the conductive disc. The mode of operation is called a single-pole generator. In figure 1–3A a fixed-level conductor is moved through a field. The motion produces a constant level of voltage across the conductor. In figure 1–3B, the conductor is rotated in the

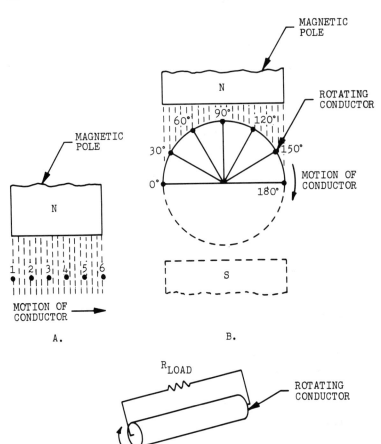

Figure 1-3 Electromechanical Generation of Electricity

magnetic field, thereby producing an alternating voltage. The magnitude of the voltage generated (E) is dependent on the flux density (B) of the magnet, the length of the conductor (L), and the speed of the conductor (V).

$$E = BLV$$

Faraday's rotational conductor was somewhat different. An equation was developed as a result of the Faraday law of induction. The induced voltage (V) produced in a conductor increases as the number of conductors (N) moving through the field increases:

$$V \propto N \frac{d\phi}{dt}$$

where V = induced voltage

$\propto$ = proportional to

$\frac{d\phi}{dt}$ = the rate at which the magnetic flux crosses the conductor

The equation suggests that induced voltage increases whenever the number of poles increases or the rate of motion of the conductor through the magnetic field increases.

The reader will note that the Faraday law of induced voltage changes somewhat with a loaded circuit. That is, when a load resistance is added to the conductor as in figure 1-3C, a current is induced through the conductor and load which opposes the motion of the conductor in the magnetic field. Without the load resistance, the conductor moves easily through the field. With the load resistance installed, the conductor does not move easily. The reason for this is that the current through the conductor produces a magnetic field which opposes the motion that generated the voltage in the first place. *Lenz's law* modifies the Faraday equation and can be stated mathematically as:

$$V = -N \frac{d\phi}{dt}$$

Note the similarity to the induced voltage equation previously stated. The minus ($-$) sign denotes the opposition developed by the load resistance.

Photoelectrical Element

The effect of light on conductive material produces an effect called the *photoelectric effect*. The sensing of light is also called *photodetection*. Photodetection is defined around three phenomena. These are *photoemission, photoconduction,* and *photovoltaic actions*. All quantum photodetectors respond directly to the action of incident light. The first, photoemission, involves incident light which frees electrons from a detector's surface (see figure 1-4). This usually occurs in a vacuum tube. Note that electron current flows from negative to positive. With photoconduction, the incident light on a photosensitive material causes the material to alter its conduction (see figure 1-5). The third phenomenon, photovoltaic action, generates a voltage when light strikes the sensitive material of the photodetector (see figure 1-6). Note that there is no power source such as a battery involved with photovoltaic action.

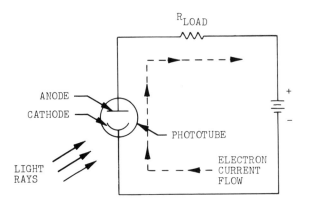

NOTE: CATHODE IS COATED WITH RADIATION
        SENSITIVE METAL. ELECTRONS ARE
        RELEASED AS RADIATION IS APPLIED.

Figure 1-4   Photoemission

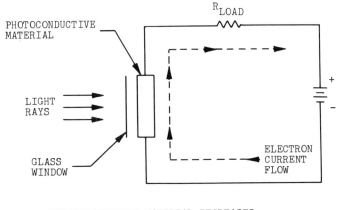

NOTE: PHOTOCONDUCTIVE MATERIAL DECREASES
        RESISTANCE AS RADIATION IS APPLIED.

Figure 1-5   Photoconduction

Piezoelectrical Element

Synthetic or natural crystals such as quartz have special physical qualities.
When crystalline material is stressed (pressure causes the crystal shape to
change or distort), a voltage is produced at the surface. If a material performs
in this manner, it is said to be *piezoelectric*. Piezoelectric material also has
another quality that is the direct reverse of piezoelectric. This other quality is

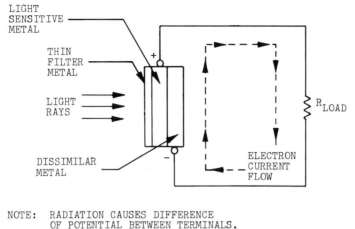

LIGHT
SENSITIVE
METAL

THIN
FILTER
METAL

LIGHT
RAYS

DISSIMILAR
METAL

+

−

R_{LOAD}

ELECTRON
CURRENT
FLOW

NOTE: RADIATION CAUSES DIFFERENCE
OF POTENTIAL BETWEEN TERMINALS.

Figure 1-6 Photovoltaic Action

called *electrostriction*. Electrostriction involves the application of an electrical field to the crystal substance. The electrical field alters the crystal shape. Probably the most common example of piezoelectric material is the ceramic material in the stylus of a disc record playback. As the record turns, the ceramic crystal flexes in the grooves, causing a voltage that represents the amount of flex. The voltage is then amplified and sent to a loudspeaker. Another common use of the piezoelectric effect is the control of radio frequencies in transmitters.

Thermoelectric Element

The sensing of temperature is usually accomplished with the aid of a thermoelectric device such as the thermocouple, the resistance temperature detector (RTD), or the thermistor. Whichever device is used, it plays a major role in the monitoring of heat energy such as in conversions from coal to steam and air conditioning.

There are three thermoelectric effects that play a large part in generation of electricity by way of temperature.

The first is the *Seebeck effect*. Thomas Seebeck, a German physicist, fused two dissimilar metal wires together on both of their ends. He then heated one of the junctions and found that electrical current flowed from one wire to the other. He caused electrons to flow from a copper wire to an iron wire. This effect developed into what we know now as the *thermocouple*.

The second thermoelectric effect is called the *Peltier effect*. Jean Peltier, a French physicist, applied current to two dissimilar metal wires attached together at their ends. As electrons moved from the copper wire to the iron wire that junction became warm (hot). As the electrons moved from the iron wire to the copper wire, that junction became cool. The reasoning is that electrons moving from a higher energy state (iron) to a lower energy state (copper) create an excess of energy, therefore heat.

The third thermoelectric effect is called the *Faraday effect*. Michael Faraday found through experimentation that in certain materials resistance is decreased as temperature increases—that is, they have a negative temperature coefficient. The resistance of oxides such as cobalt, manganese, and nickel is decreased when they are heated.

Thermoelectric materials are chosen for their ability to provide a uniform voltage-temperature relationship.

PASSIVE TRANSDUCER ELEMENTS

Each passive transducer has an element which, under some force, responds by moving mechanically to cause an electrical change. Lets's discuss some of these elements in a cursory manner.

The Capacitive Element (See figure 1-7)

The capacitor consists of a pair of plates made of conducting material placed each side of or around a nonconducting material (insulator) known as a *dielectric* (see figure 1-7A). Leads from each conducting plate are connected to the circuit. The capacitor element operates in the same manner as a simple capacitor (see figure 1-7B). Force from a measurand (that which is being measured) moves one or both of the plates, which changes the capacitance of the capacitor. In an excited circuit, the capacitance change would cause a change in capacitive reactance which, in turn, would modify current flow.

Consider what may happen in the event of a measurand change. The exact formula for calculating the capacitance of a capacitor depends upon the capacitor's size, shape, and dielectric constant. The basic formula for calculating capacitance of a parallel plate capacitor is:

$$C = \frac{kA}{D}$$

where C = capacitance in farads

k = dielectric constant

A = area of either plate

D = distance between the plates

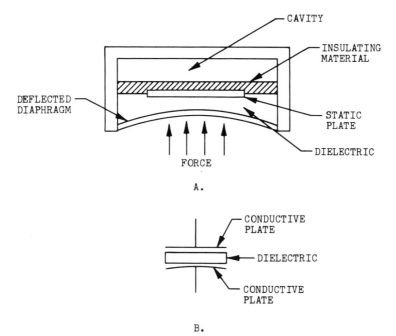

Figure 1-7   The Capacitive Transducer

The formula may be expanded in the following manner:

$$C = \frac{kA}{D} \times 22.4 \times 10^{-14}$$

where the value 22.4×10^{-14} farad is used to convert the C to a farad unit

The formula indicates that any of the physical factors that are involved in the capacitor's makeup could be used to change its capacitance.

The dielectric constant is dependent on the material from which the dielectric is made. Typical of the dielectrics are vacuum, paper, mica, ceramics, and glass. Vacuum is the base dielectric constant (1), while other examples are air (1), paper (2), mica (3), glass (8), and ceramics (100).

Another equation that is important to the capacitive transducer is the quantity of charge (Q) of electricity that can be stored in the capacitor. The formula is:

$$Q = CV$$

where   Q = quantity of charge in coulombs

C = capacitance of the capacitor

V = voltage applied in volts

The charge on a capacitor is made by repelling free electrons from one plate of the capacitor (+) and attracting them to the opposing plate (−). This causes a difference of potential across the dielectric equal to the applied voltage. To discharge a capacitor, a short is placed across the dielectric. The action allows free electrons to move back to atoms that have open holes in their valence rings, thereby stabilizing the atoms.

The time to charge a capacitor is important in capacitive circuits and is used as a parameter in transducer applications. In general, the capacitor is used along with circuit resistance to determine the time constant of the circuit.

The time constant in a resistive-capacitive (RC) circuit is the time in seconds that it takes for the capacitor in the circuit to charge to 63.2 percent.

An RC circuit contains at least one capacitor and one resistor. Figure 1–8A illustrates this circuit with a voltage input of 5 volts. When the switch is closed, current begins to charge the capacitor. The capacitor charges by 63.2 percent of the applied voltage, as follows:

Applied Voltage	5.00 volts
1st Time Constant	− 3.16 volts
Remaining Voltage	1.84 volts

During the second time constant, voltage will increase 63.2 percent of 1.84 volts: $0.632 \times 1.84 = 1.16$ volts increase.

The charge voltage on third, fourth, and fifth time constants increase in the same manner. After the fifth time constant, for practical application, the charge has reached the full 5.0 volts applied.

While the charge is building up from 0 to 5 volts, the capacitor is opposing a change in voltage. The reaction of the RC circuit during this time is called *transient response.* The time constant (TC) in seconds is a ratio of capacitance and resistance.

Figures 1–8B and 1–8C are curves that represent charge and discharge time, respectively:

$$TC = RC$$
$$TC = 5000 \times 0.0001 \ (5 \ k\Omega \times 100 \ \mu F)$$

If the closed circuit shown in figure 1–8 is now opened, the reaction of the circuit will be that the capacitor will discharge through the resistor by 63.2 percent during the first time constant, then by 63.2 percent of the remaining charge during the second time constant. The capacitor will continue to discharge in the same manner, and after the fifth time constant will, practically speaking, have a

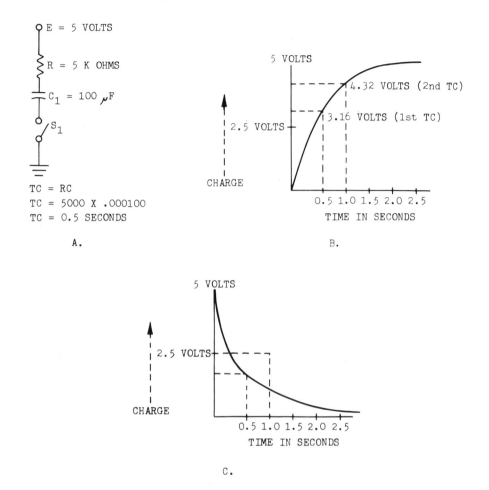

Figure 1-8 RC Circuit Time Constant

zero charge. The time constant grows in value if either the capacitance or the resistance is increased. Conversely, the time constant becomes less if either the capacitance or the resistance is decreased.

The Inductive Element (See figure 1-9A)

The inductive transducer element consists of a diaphragm or core called an *armature* that is driven by force from a measurand. The armature is either displaced or rotated near the coils of a C-shaped pickoff. The displacement or rotation of the armature causes a change of inductance of the magnetic flux in the coil by varying the air gap in the flux path. In an excited circuit this would

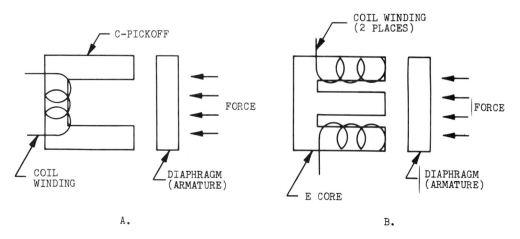

Figure 1-9 The Inductive Transducer

cause a change in inductive reactance which, in turn, would modify current flow.

In figure 1-9B the armature is either displaced or rotated between a pair of coils of an E-type pickoff. The displacement or rotation of the armature causes a change of inductance between the two coils.

Let's consider what may happen in the event of a measurand change. The parameters that are involved in the physical makeup of the coil can be modified by the force of the measurand. The inductance of a coil or inductor may be changed by the core material, the relationship between the armature and the E-pickoff, the number of coil turns, the diameter of coil, and the coil length.

An inductor operates on the principle of induction. Current through an inductor lags voltage by 90° in a pure inductive circuit and by some angle between 0° and 90° in other RL circuits. As current increases through an inductor, a magnetic field is built up around the coil. The strength of this field is dependent on the physical makeup of the inductor and the core permeability. The magnetic field produces a counterelectromotive force which opposes a change in current.

The core of the inductive pickup has some properties that cause some error. The most prominent of these losses are in *eddy currents* and *hysteresis*. Eddy currents are prevalent in iron cores. Alternating current induces voltage in the core, causing currents to flow in a circular path through the core. This causes a loss in signal power. Eddy currents occur more often at high-frequency ac operation. The second type of loss is hysteresis. Hysteresis also occurs most prominently at high frequencies; it is loss of power as a result of switching or reversing the magnetic field in magnetic materials.

The Q of a coil represents its ability to store internal energy. The factor is a ratio between the inductive reactance of the coil and its internal resistance:

$$Q = \frac{X_L}{R_L}$$

where Q = quality or merit of the coil
X_L = inductive reactance of the coil R_L
R_L = internal resistance of the coil

At low frequencies, the internal resistance of the coil is purely the dc resistance of the wire. At high frequencies, power losses from eddy currents increase and the Q of the coil decreases. Q of a coil can be identified by the ratio of reactive power to resistive or real power:

$$Q = \frac{P_{XL}}{P_R}$$

where Q = quality or merit of the coil
P_{XL} = reactive power
P_R = resistive power

Significantly, the time that it takes for the field to build up and collapse in an inductive circuit is an important factor in transducer applications. In general, the inductor may be used as a timing component along with the resistive elements in a circuit.

The time constant in a resistive-inductive circuit is actually the time that it takes for the current to change by 63.2 percent. An RL circuit contains at least one inductor and one resistor. Figure 1–10A illustrates this circuit, which has 1.0 ampere of steady-state current flow.

When the switch is closed, current starts to flow, and in one time constant increases from 0 to 0.632 ampere. (See figure 1–10B for a curve that represents the field being built up and the relationship between curve and time.) During the second time constant, current increases by 63.2 percent of the remaining steady-state current, as shown below:

Steady-State Current   1.000
1st Time Constant   − 0.632
Remaining   0.368 ampere
63.2% of 0.368 = 0.232 ampere increase
during 2nd time constant

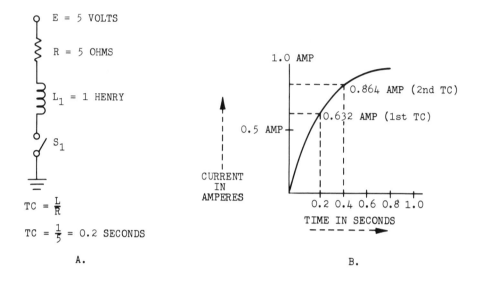

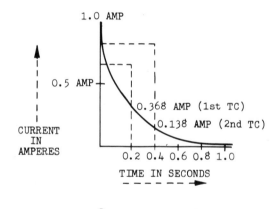

Figure 1-10   RL Circuit Time Constant

Current continues to increase for the third, fourth, and fifth time constant in the same manner. After the fifth time constant, for practical application, current has reached the 1.0 ampere steady-state current.

    While current is building up from 0 to 1.0 ampere, the inductor is opposing the change in current. The reaction of the RL circuit during this time is called *transient response*. The time constant (TC) in seconds is a ratio of inductance and resistance, as follows:

$$TC = \frac{L}{R}$$
$$TC = \frac{1 \text{ henry}}{5 \text{ ohms}}$$
$$TC = 0.2 \text{ second}$$

If the closed circuit shown in figure 1–10A is now opened, the reaction of the circuit will be that the current will decay 63.2 percent in the first time constant, then by 63.2 percent of the remaining current in the second time constant.

The current will continue to decay in the same manner, and after the fifth time constant will, practically speaking, be zero. (See figure 1–10C for a curve that represents the field collapsing and the relationship between current and time.)

The time constant grows larger with an increase in inductance and smaller with an increase in resistance.

The Potentiometric Element (See figure 1-11)

Resistive transducers may be potentiometric. There are available other resistive elements such as the slide resistor or other forms of variable resistance. A *potentiometric transducer* is an electromechanical device consisting of a resistive element with a movable wiper or slider. A measurand causes a force to act on the wiper. The wiper makes contact along the resistance in relation to the amount of force applied by the measurand. As the wiper moves, the output may be taken from between one end of the resistance and the wiper. Output values may be linear, trigonometric, logarithmic, or exponential.

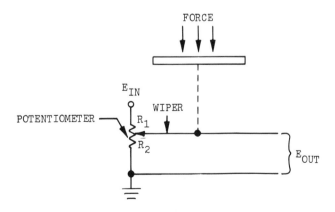

Figure 1-11   The Potentiometric Transducer

The potentiometric transducer is usually large in size, although there has been considerable progress in miniaturization. The potentiometric transducer also has high friction problems and is extremely sensitive to vibration. It develops high noise as it ages and has low frequency response. Despite these problems, the potentiometric transducer is inexpensive as compared to other transducers. The device can be easily installed and can be excited by both alternating (ac) and direct (dc) current. Finally, the potentiometric transducer has a high output with no amplification or impedance matching problems.

The basic formula involved with the potentiometric transducer in and out voltages is as follows:

$$E_{OUT} = E_{IN} \frac{R_2}{R_1 + R_2}$$

This is a voltage divider formula whose output is taken from R_2, the lower side resistance as picked off by the wiper.

If the output was taken from resistance R_1 (the upper side of the wiper), the polarity of the output would change ends and the voltage divider formula would change slightly:

$$E_{OUT} = E_{IN} \frac{R_1}{R_1 + R_2}$$

It should be obvious to the reader that there are certain factors that will affect the output of the potentiometric or resistive transducers. One of the more important is the resistance of the wire. Wire-wound resistance and most potentiometers are similar in structure. Wire itself is usually measured in American Wire Gage (AWG). As an example, AWG 16 is 50.8 mils in diameter, 2581 circular mils in area, and measures 4 ohms in resistance. The diameter in this example is 50.8 mils. To convert diameter in mils to circular mils, one must simply square the diameter in mils ($50.8 \times 50.8 = 2580.64$). You should note that this is not an accurate figure, but only a choice by standards to create a useable value. The circular mils value is actually the area of a square. To convert this value to the area of the circular cross-section of a wire, the following formula for a circle should be used:

$$A = \frac{\pi D^2}{4}$$

where A = area of the circular cross-section

π = 3.1416

D = diameter of the wire in mils

The material from which the resistor or potentiometer is made becomes a factor in the realm of effects. The material has a resistivity factor developed by American Standards for each material. For instance, copper has a resistivity of 10.37 ohms per foot length with a cross-section of 1 mil in area. This is a mil-foot. Silver has a resistivity of 9.9 ohms per mil-foot, gold 14 ohms per mil-foot, and tungsten 33 ohms per mil-foot. These values represent the ability to resist current flow.

A practical relationship to determine the resistance of a resistor, potentiometer, or in fact a piece of wire, is calculated by the following equation:

$$R = \frac{kL}{A}$$

where R = resistance in ohms
   k = resistivity of the conductor
   L = length of the conductor
   A = area of the conductor

It should be obvious that any change in length, area, or resistivity will change the total resistance. These change factors, of course, can be useful in developing transducers.

As with other transducers, temperature always is a factor that effects resistance change. Some materials have positive temperature coefficients, others have negative temperature coefficients. The positive coefficients relate to a resistance increase with an increase in temperature. The negative coefficients relate to a decrease in resistance with an increase in temperature. A basic formula is also available to consider the amount of resistance change due to temperature. Thus:

$$\Delta R = \alpha_t \, R_i \, \Delta T$$

where ΔR = change in resistance
   α_t = temperature coefficient of the material
   R_i = initial resistance of the material
   ΔT = temperature change

Thermistors have been manufactured for many years. They are made to increase or decrease in resistance purposefully as temperature increases or decreases. With devices such as the thermistor, the amount of change is significant. A very small temperature change may cause a very large and accurately predicted change in resistance. This provides the designer with a useful parameter in the development of a transducer or a system.

SOME SPECIAL TRANSDUCER ELEMENTS

Some transducers employ the basic action of active or passive transducers. Others may employ some special combination of characteristics that set it partly aside from the standards determining active or passive qualities. We will examine some of these transducers.

Electrokinetic Elements (See figure 1-12)

An electrokinetic transducer consists of polar fluid contained between a pair of diaphragms. A porous disc is inserted in the polar fluid between two partially porous plates called *diaphragms*. A change in force from the measurand causes the diaphragm to deflect and allow a very small amount of polar fluid to flow through the porous disc partition to effect a deflection on the second diaphragm (left in the illustration). This flow causes a difference of potential between the plates of the porous plug. A second effect may be recognized in reverse operation. An electrical potential may be applied to plates of the porous plug, which in turn causes polar fluid flow and diaphragm deflection.

The electrokinetic transducer element is self-generating with relatively high-frequency response and high output. It does have disadvantages in that it cannot monitor static pressures or linear accelerations. The element cannot be calibrated without flow. A further disadvantage is that the polar fluid is often volatile.

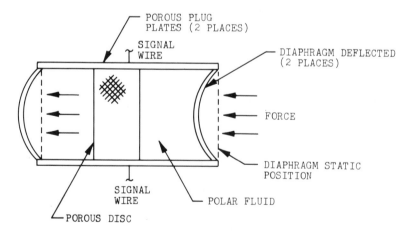

Figure 1-12 Electrokinetic Transducer Element

Force Balance Element (See figure 1-13)

In the figure, the actual sensing element is the capacitor in the force balance element. However, this element could just as well be an inductive element. Operation of this device is rather simple. A change is effected by a measurand change and force is applied to a force summing capacitor. The physical properties of the capacitor change and therefore its capacitance also changes. The variable signal is routed to an amplifier and in turn to a servo mechanism. A feedback signal equal to the variable output of the capacitor is fed back to the force balance element to return it to its original state before a measurand force change. The feedback may be mechanical and/or electrical and may be mechanized by a servo system or electromagnetism. Actual motion of the mechanical linkage and force summing devices is hard to detect. The device has a fairly high output and is accurate and stable. It does, however, have a low-frequency response and is sensitive to acceleration and shock. The element is heavy and may be expensive.

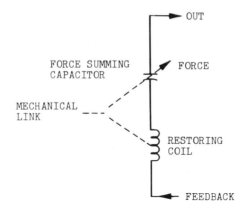

Figure 1-13   Force Balance Transducer

Oscillator Element (See figure 1-14)

In the figure, a fixed capacitor and a variable inductor (coil) are used as a frequency-resonant circuit input to an oscillator. The variable unit may be either of these components. The force element may be such as water or air pressure. As the measurand changes, the force summing bar causes the inductor to change inductance, which alters the frequency at which the circuit is reso-

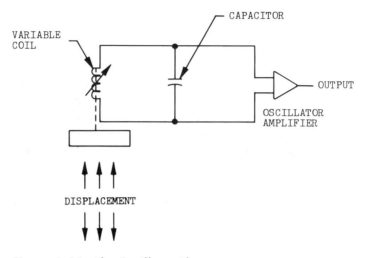

Figure 1-14   The Oscillator Element

nant. Any change in the resonant circuit will be felt at the oscillator amplifier input:

$$\Delta f = f_o - f_r$$

where Δf = change in frequency

f_o = operating frequency

f_r = resonant frequency

The oscillator element may be very small with a fairly high output. Temperature may restrict the operating range and the devices used may play a part in poor sensitivity. Accuracy of the device may be poor unless expensive components are utilized.

Differential Transformer Element (See figure 1-15)

The differential transformer consists of primary coil, a pair of or several secondary windings, and a core. The primary and secondary windings are inductors (coils) and, as you would expect, have properties similar to the inductive transducers we have just discussed. The core is moved by a force from the measurand. In a majority of designs the core mass is a push rod of some description which is tied to linkage. The core moves between the primary and secondary coils as the measurand changes. Induced voltage between the primary and secondaries changes and an output is effected. The secondary output is usually fed to a demodulator or an ac bridge network.

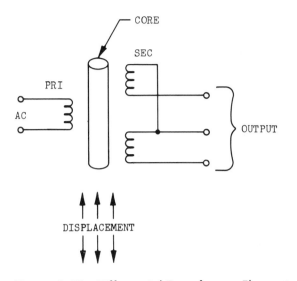

Figure 1-15 Differential Transformer Element

The differential transformer transducer is best suited for large displacements. Signal output is larger than for most other transducers. The differential transformer is always driven by alternating current. The frequency of the ac current is usually 50 to 60 hertz from commercial power sources. Higher frequencies of 8kHz or 10kHz may be utilized to reduce sensor component sizes. Because of its large displacement, the differential transformer is susceptible to vibration. However, some of the modern linear models have overcome this problem.

Again, all the factors that affect inductors and inductive transducers will also affect the differential transformer.

Photoelectric Element (See figure 1-16)

The photoelectric element consists of a force summing diaphragm, a window, a light source, and a detector. Some measurand such as force or pressure causes the diaphragm to deflect. The window opens or closes as the diaphragm modulates the amount of light from the light source through the window opening. A photodiode on the window's opposite side detects the light and feeds the varying amounts as current flow to some receiving or amplifying component. The output then is dependent on the force applied by the measurand. This is a linear system and can be made extremely accurate with a strong and sensitive light-emitting diode (LED) as the light source. The unit is simple and can provide a high output.

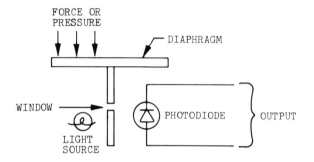

Figure 1-16 The Photoelectric Element

Frequency response may be low and long-term stability is not always achieved. Temperature, as in other transducer elements, is always a factor in efficiency.

Vibrating-Wire Element (See figure 1-17)

The vibrating-wire element consists of a fine tungsten wire strung through a strong magnetic field so that the lines of force crossing the wire are maximum. As the measurand causes a force change, the wire vibrates at a frequency that is determined by the length of the wire and the tension applied. As the wire vibrates in the magnetic field, an electrical signal is generated at the output (taken from the wire) and fed to an amplifier. A feedback signal is fed back to the wire to maintain oscillation at the desired frequency. Modification of the wire-vibration frequency is then determined by the force applied. The output of the vibrating-wire element can be high and can be transmitted over long distances without much loss. The device is sensitive to acceleration and shock and is not considered a stable element. Temperature affects the wire and its hookup.

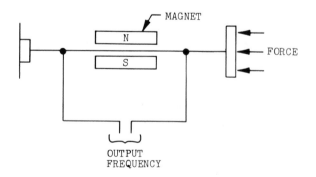

Figure 1-17 The Vibrating-Wire Element

The Velocity Element (See figure 1–18)

The velocity element consists of a moving coil within a magnetic field. One end of the coil is attached to a pivot while the other end is free to move within a restricted area. Output is taken from the moving coil. The voltage signal is generated by the coil moving within the magnetic field. The output is proportional to the velocity of coil movement. Electrical feedback may be used for dampening purposes.

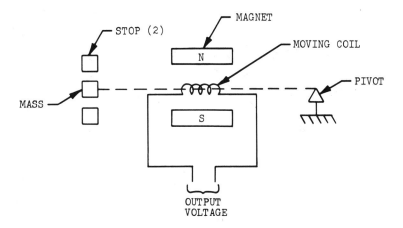

Figure 1–18 The Velocity Element

The Strain-Gage Element (See figure 1–19)

The purpose of a strain-gage element is to detect the amount or length displaced by a force member. The strain gage produces a change in resistance which is proportional to this variation in length. Each strain gage has a property known as a *gage factor* which provides this function. Gage factor is a ratio of resistance and length:

$$GF = \frac{\Delta R/R}{\Delta L/L}$$

where GF = gage factor

ΔR = change in resistance

R = original resistance

ΔL = change in length

L = original length

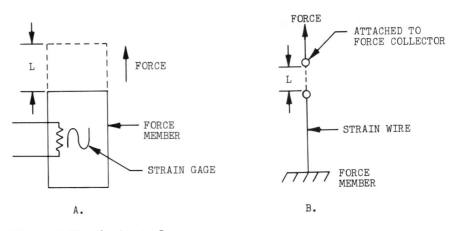

Figure 1-19 The Strain Gage

The gages are electrically installed as part of a Wheatstone bridge for circuit applications.

There are two basic strain-gage types: the *bonded* and the *unbonded*. The bonded gage (figure 1-19A) is entirely attached to the force member by adhesive. As the force member stretches in length, the strain gage also lengthens. The unbonded gage (figure 1-19B) has one end of its strain wire attached to the force member and the other end attached to a force collector. As the force member stretches, the strain wire also changes in length. Each motion of length either with bonded or unbonded gages causes a change in resistance. Strain gages are made of metal and semiconductor materials. Strain gages are very accurate, may be excited by alternating or direct current, and have excellent static and dynamic response. The signal out is small, but this disadvantage may be corrected with good periphery equipment.

Other Transducers There are certainly other specialized transducers made in this country and the rest of the world that do special tasks and have unique functions. Throughout this book, we will attempt to provide at least cursory coverage of most of today's transducer types.

NATURAL FREQUENCY

Most transducers that are involved with static acceleration or vibration have what is called an *expected natural frequency*. The natural frequency of a seismic mass possessing one degree of freedom is as follows:

$$f_n = \frac{1}{2\pi} \sqrt{\frac{K}{M}}$$

where f_n = natural frequency

K = maximum stiffness

M = minimum mass

Sensitivity of acceleration and vibration transducer devices increase in proportion to mass (M) and decrease in inverse proportion to stiffness (K).

Transducers that do respond to vibration or acceleration have a particular point at which they freely oscillate without being forced. This results in some transducers responding with a ringing noise. Ringing may also occur with a step change in the measurand. Others simply vibrate or oscillate at their own frequency. Manufacturers will freely provide information regarding the transducer's natural frequency and/or resonant points.

ERROR

A measurement evolves around three separate concept parts. These concepts are the *quantity or quality of the physical phenomena* being measured, the *reference for comparison,* and the *method of comparing sensed quantity or quality* to the reference.

Likewise, the act of measurement must be based on three fundamental operations: sensing, processing, and utilizing collected information.

Regardless of the components used in the measurement, there is room for error. Error is usually considered to be static or dynamic. *Static error* can be further separated into predictable errors such as parallax, interpolation, and environmental causes such as heat, humidity, pressure, and radiation. *Dynamic error* is that difference between the sensed physical quantity and the value indicated after installing the measuring equipment. In other words, dynamic error is a result of loading or connecting the measurement system.

Error must be determined and either automatically compensated for by the system or added or subtracted from indicated results.

MEASUREMENT

Measurement, according to Webster, is a broad subject. First, it has to do with measuring or mensuration. Second, measurement is determining the extent, quantity, or size of something. Third, measurement is also *a system of* measuring or of measures. It is really not necessary to separate these three, for in the field of transducers, we use them all.

MEASUREMENT PROCESSES

A forward thought in the world of testing and measurement was provided by Robert W. Lally of PCB Piezotronics, Incorporated. In his article "Metrigenesis" in *Test* magazine (Aug./Sept. 1978), he wrote of a natural concept in the field of measurement. The word "metrigenesis" was developed from the root *metri,* meaning measurement, and *genesis,* meaning ongoing origin. Thus "metrigenesis" describes the involutionary process of testing and measurement.

Measuring both stimulus and response is common practice in testing the behavior of things. Today, at an accelerating pace, we are witnessing, experiencing, and intimately involved in the development and use of measuring instruments.

Transduction employs an energy transfer process to sense and communicate information (see figure 1–20). When a structural model transfers energy, involves a dual transaction, and manifests some physical law or effect, it is called a *transducer.* When a transducer primarily functions to sense and transfer information into a more convenient form, it is a *measuring transducer or sensor.* And, depending upon its location in the structure of a system, a model can be an *input, modifying* (conditioning), or *output* transducer. In the measuring transaction, sensor structures obeying the laws of nature also tend to delay, distort, and degrade the information being communicated and change the quantity being measured.

Since a measuring transaction involves a reaction and an energy transfer, it will always in some small way change the quantity being measured, and thereby affect the validity of the measurement. The purpose of the measurement—research, test, control, or calibration—imposes different tolerances on

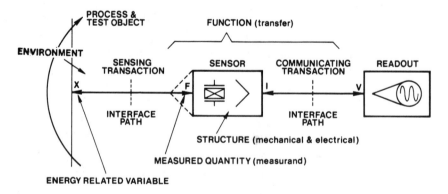

Figure 1–20   Transduction Process *(Courtesy, PCB Piezotronics, Inc.)*

this interaction effect. *Calibration* is testing the transfer behavior of a sensor in controlled transactions.

As illustrated in figure 1–21, measuring systems are composed of elements arranged in an organized structure and physically separated into component blocks. This model, patterned after human behavior, includes both the basic measuring process and related activities. The feedback path represents a basic mode of human behavior wherein the actual (measured) condition is compared with a desired (normal or predicted) condition and corrective, controlling, or adaptive action is taken according to the difference. For this process, one needs *norms and standards.*

The concept of test and measurement has many practical ramifications. A few examples are given in the following material.

The measurement process involves *sensor systems* (see figure 1–22). Modeled after human nature, sensors (or transducers, as they are often called) relate to human senses, transferring environmental stimuli into electrical signals. Cables relate to human nerve fibers, communicating information to the brain. The readout instrument (e.g., oscilloscope), like the mind, forms a picture or image of the sensed phenomena. To complete the feedback process, the

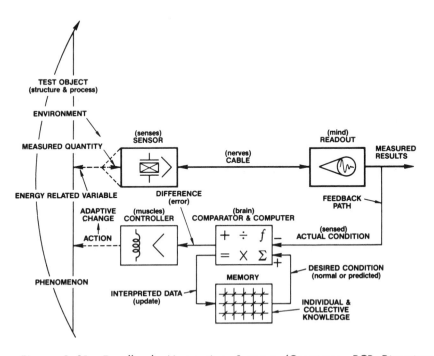

Figure 1–21 Feedback Measuring System *(Courtesy, PCB Piezotronics, Inc.)*

ABINGDON COLLEGE OF FURTHER EDUCATION

Library

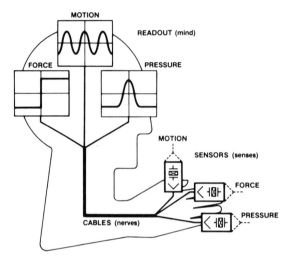

Figure 1-22   Sensor Systems *(Courtesy, PCB Piezotronics, Inc.)*

sensed picture is compared with a standard, expected, or desired one stored in the memory. Corrective or controlling action is taken according to the difference between sensed and expected values.

PHYSICAL MEASUREMENT PARAMETERS FOR TRANSDUCERS

Table 1-2 lists the most common physical measurement parameters in use along with their symbols, SI units, and transduction principle. The symbols provided are the preferred symbols used in most physics and engineering activities.

TABLE 1-2 Physical Measurement Parameters

MEASUREMENT PARAMETER	SYMBOL	S.I. UNITS (ENGLISH UNITS)	TRANSDUCTION PRINCIPLE
LINEAR DISTANCE	l	meters, m	Capacitance, incremental or differential
(Displacement)	X	centimeters, cm	Inductive, eddy current, variable reluctance
(Dimension)	S	millimeters, mm	
(Position)		micrometers, μm	Strain gage, bonded, or semiconductor
		(inches, in.)	
		(feet, ft.)	Linear Voltage Differential Transformer (LVDT)
			Optical, photo electric, or Laser Interferometer
			Linear Potentiometer
			Linear (Digital) Encoders

TABLE 1-2 *continued*

MEASUREMENT PARAMETER	SYMBOL	S.I. UNITS (ENGLISH UNITS)	TRANSDUCTION PRINCIPLE
LINEAR VELOCITY Speed	$\dot{X}$ V $\dot{S}$	meters per second m/s kilometers per second	Inertial Mass—Magnetic field—self generating Pendulous Mass—Spring—LVDT —Potentiometer Inductive—attached linkage —non-contact Optical—time differential Piezoelectric—integrated acceleration
LINEAR ACCELERATION	a g $\ddot{X}$	meters per second per second m/s² (feet per second per second. F/S²)	Seismic Mass—piezoelectric —piezoresistive —strain gage —LVDT —inductive —capacitance —potentiometer Force balance servo
ANGULAR DISPLACEMENT	$\ominus$	radians, r (degrees) revolutions (cycles) (arc-seconds)	Capacitor Inductive Resistor Photo electric Strain gage LVDT (rotary) Gyroscope Shaft Encoder, digital
Inclinometer	$\ominus$	radians (grade) (degrees)	Pendulum—Potentiometer Force balance—accelerometer—integrator Liquid—resistor
ANGULAR VELOCITY (Tachometer)	$\ominus$	radians per second r/s (rpm)	Generator, dc , ac , drag cup Photo electric or magnetic pulse wheel
ANGULAR ACCELERATION	$\ddot{\ominus}$	radians per second per second r/s²	Electromagnetic + differentiator Force balance servo Digital Shaft Encoder

(cont.)

TABLE 1-2 *continued*

MEASUREMENT PARAMETER	SYMBOL	S.I. UNITS (ENGLISH UNITS)	TRANSDUCTION PRINCIPLE

FORCE
$F = ma$
	F	Newton, N
		Kilogram-force
		dynes
		
		(lb-f)
			Counter balance—Mass
			       —electromagnetic
			
			Deflection—Strain Gage
			   —LVDT
			   —Piezoresistive
			   —Capacitive
			   —Inductive
			   —piezoelectric

TORQUE	T	Newton-Meter
		$N \times m$
		dyne-cm
		(lb-ft)
		(oz-in)
			Torsional Windup—Strain Gage
			   —LVDT (Torsional Variable D.T.)
			   —Photo-electric encoder
			   —Permeability change
			Dynamometer

VIBRATION—DISPLACEMENT (Distance) (Amplitude)	D peak to peak d peak	millimeters, mm (inches, in)	Linear Displacement Transducers (DC-LVDT) Integrated linear velocity transducer signals Double integrated accelerometer signals
VIBRATION—VELOCITY	V zero to peak	millimeters per second mm/s (inches per second) (ips)	Linear velocity transducers, seismic Integrated accelerometer signals
VIBRATION—ACCELERATION	G, g, a	g's meters per second per second	Linear accelerometers-piezoelectric
VIBRATION-TORSIONAL			Time interval variations during rotation of each revolution. Strain gage—inertial mass
SOUND (Microphone)		dBm dBv	Piezoelectric Capacitance Variable reluctance

TABLE 1–2 *continued*

MEASUREMENT PARAMETER	SYMBOL	S.I. UNITS (ALTERNATE UNITS)	TRANSDUCER DESCRIPTION
PRESSURE	Pa, p	Pascal Newton per square meter N/m² (psi) (mm Hg)	Bellows—Potentiometer Capsule—Differential transformer—(LVDT) Diaphragm—Strain Gage Diaphragm—Piezoresistive Diaphragm—Piezoelectric Diaphragm—Variable Capacitance Diaphragm—Variable Inductance
FLOW METERS		Cubic meter per second m³/sec (sc cm) (ft³/min) (cfm) (gpm) (ccm) Kilogram per second Kg/s, (SCCM) (SCFM) (lb/min) (lb/hr)	Positive Displacement—Volumetric—liquid Positive Displacement—Volumetric—gas Differential Pressure—orifice $Q = K\sqrt{\dfrac{P2}{P1}}$ —Venturi —Pitot tube Turbine—velocity—liquid —gas Magnetic—velocity Variable area—float meter —force meter Thermal—mass flow Differential Pressure—mass flow Turbine—axial—momentum—mass flow

TEMPERATURE		Degrees Celsius
		°C
		(°F)
		(°K)
		(°R)
			Thermocouple—K—Chromel-Alumel
			    J—Iron-Constantan
			    B—Platinum-Rhodium
			    T—Copper-Constantan
			RTD—Platinum
			  —Nickel
			  —Thermister
			
			Semiconductor junction
			
			Pyrometer—radiation
			  —Optical

(*cont.*)

Table 1-2 *continued*

MEASUREMENT PARAMETER	SYMBOL	S.I. UNITS (ALTERNATE UNITS)	TRANSDUCER DESCRIPTION
VISCOSITY Absolute-poise Dynamic-stoke	$\eta, (\mu)$	Newton-second per square meter N-s/m² (dyn-sec/cm²) centipoise, cp	Falling—Ball —Piston Capillary or Orifice (Saybolt) Rotating member
HUMIDITY		Relative Humidity R.H. Parts per million ppm	Animal hair—mechanical linkage Lithium chloride, electrical resistance Capacitance Microwave Thermoelectric—optical servo

The SI units are the Standard International Metric units. These are listed first, while English units are shown in brackets. More common nomenclature is abbreviated. The physical and electrical sensing form is listed for the most-used transducer. A more generalized term is used instead of trade names. No order of preference is attempted. All items listed are generally available and represent the majority of installed units.

REFERENCES

Reference data and illustrations in Chapter 1 were supplied by state-of-the-art manufacturers. Permission to reprint was given by the following companies:

PCB Piezotronics, Buffalo, New York
Physical measurement parameters—Tektronix, Beaverton, Oregon
All copyrights © are reserved.

2

Transducer Motion Mechanization

A very large number of transducers are involved in general motion, displacement from a fixed point, or at least in a form of position as it relates to some reference. These words all may be synonymous. The three dominant transducer motion mechanization components are the *potentiometer,* the *synchro,* and the *linear variable differential transformer* (LVDT).

All three are electromechanical devices designed to provide an output directly related to some form of motion. This chapter is dedicated to these devices.

THE POTENTIOMETER + P40/41
OTHER BOOK

A Basic Potentiometer A potentiometer, formed by a slider that is movable along a resistance element, is shown schematically in figure 2–1A. A more familiar representation, with excitation and output voltages added, is depicted in figure 2–1B.

Although the concept of such a potentiometer is simple, the designs of actual units are sufficiently complex to warrant a general discussion of their principal elements. This section describes typical elements of single-turn linear and nonlinear potentiometers, and discusses some of the factors contributing to their accuracy. Multiturn and translatory potentiometers contain many of the basic elements of the single-turn design.

The Base

The base of a single-turn precision potentiometer, shown in figure 2–2, is a cup-shaped housing with a cylindrical wall and a central cylindrical hub for support-

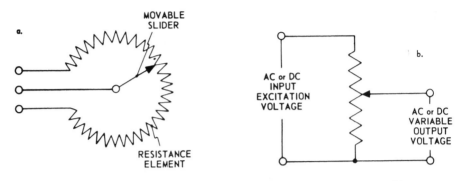

Figure 2-1 Schematic Representations of a Potentiometer (Courtesy, Bowmar/TIC Inc.)

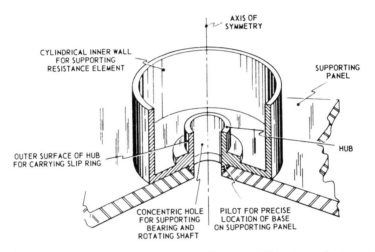

Figure 2-2 Potentiometer Base Cutaway (Courtesy, Bowmar/TIC Inc.)

ing a rotatable shaft. The function of the base is to hold the operating elements of the potentiometer in precise relation to one another and in precise and rigid alignment with a panel or other supporting structure.

The proper functioning and accuracy of the potentiometer depends on the precision of outer, as well as inner, dimensions of the base. The inside surface of the cylindrical wall, to which the resistance element is bonded, must be uniformly circular and concentric with the axis of symmetry.

Both the inner and outer surfaces of the hub require precise dimensioning so that a bearing and shaft that fit inside the hub and a slip ring, which fits over the hub, will be concentric with the cylindrical wall of the base. The device used to adapt the unit to its mount—such as a locating pin, a threaded bushing, a precision pilot, etc.—must be mechanically accurate so that the unit can be positioned within close tolerance to gears, drive motors, or similar related mechanisms on the mount.

Potentiometers are available with metal or plastic bases. Metal units are machined from castings or bar stock; plastic units are usually cast in a precision mold. The metal surfaces are finished for protection against corrosion, and the plastic surfaces are resistant to moisture absorption and fungus growth.

Where high accuracy is required, metal bases are superior, since the dimensions are more precisely controlled and more rigid. Where low costs or high-frequency operation are a consideration, plastic units are sometimes preferable, since they are less expensive and shunt capacitance between the base and the resistance winding is kept to a minimum.

Resistance Element ₊ P 17 OTHER BOOK

The resistance element is a critical part of the precision potentiometer. It is usually formed by an insulated resistance wire wound on a supporting structure. A varnish coating serves to hold the wires in place and protect them from moisture and damage. An area of the winding is cleaned to obtain a straight, smooth path over which a slider can maintain good electrical contact. Except in the case of toroidal windings wound about a solid ring, the resistance element is wound flat and then bent into circular shape to fit within the base.

The supporting structure for the winding may be a flat, thin card of flexible insulating material (usually a phenolic plastic) or a round rod, called a *mandrel*, usually of polyester resin-insulated copper wire. The phenolic card is used for both linear and nonlinear potentiometers and rheostats; the round mandrel is well adapted for multiturn potentiometers. Illustrations of card and mandrel winding supports are shown in figure 2–3.

A principal advantage of the card-type winding support is that high slope ratio functions can be generated using large resistance wire diameters.

The wound card also provides a flat contact surface where the resistance wire is not strained and the slider can travel with minimum wear. The flat surface also allows the use of formed-wire multiple-contact sliders and flat conducting resistively shorted overtravels.

The principle advantages of the rod-type mandrel are its smooth and uniform dimensions and the ease of winding such a cylindrical form. Because of this uniformity, greater accuracy is achievable than with the equivalent card-type support. Actual resolution is superior because all the turns of resistance wire are contracted. The round mandrel design usually results in a shorter mechanical package than the equivalent card-type unit.

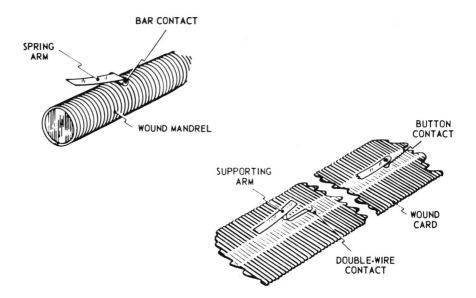

Figure 2–3 Wire-Wound Resistance Elements on Card and Mandrel Supports *(Courtesy, Bowmar/TIC Inc.)*

The function generated by the potentiometer is determined by the variation of the spacing, the shape of the supporting card, and the values of winding resistivity that are used. A few of the factors upon which the accuracy of the resistance element depends are that the dimensions of the card be precisely maintained, that the resistivity of the wire remain uniform, and that the spacing of the windings be carefully controlled. In the selection of resistance wire alloy and in the cleaning of the slider contact path, special measures can be taken to minimize the noise output of the potentiometer.

Conducting Overtravels

Each end of the resistance winding is joined to a conducting overtravel, which may take the form of a short metallic strip of negligible resistance. The overtravels are fitted and bonded to the supporting card so that the slider makes a smooth transition as it passes from the surface of the winding to the surface of the overtravel. Overtravels may also be formed by shorting together a series of turns at each end of the winding.

The overtravels provide an angular distance of about 10 degrees at both ends of the resistance element over which the slider can move without changing

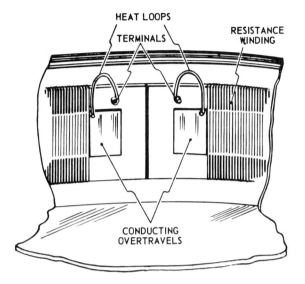

Figure 2-4 Conducting Overtravels at Ends of the Resistance Element (*Courtesy, Bowmar/TIC Inc.*)

the potentiometer output. Each overtravel is joined to an external terminal by means of a length of wire, as shown in figure 2–4. When a soldering iron is applied to the external lug, the length and shape of this wire helps to radiate the heat before it reaches the resistance winding itself.

Conductive overtravels on a mandrel are achieved by shorting the resistance wire with conductive materials. These may be epoxies, metal sprays, solders, or solid precious-metal sections.

Taps (See figure 2-5)

A tap is an electrical contact fixed at some known angular point of the resistance winding and connected through a lead to an external potentiometer terminal or to the internal shunting resistor.

Taps are made by welding a resistance wire to one turn of the winding. These taps are located angularly or electrically. A tap made to a .001-inch diameter mandrel will carry 200 MA of current and the base may be picked up by this tap wire. Usually the tap wire is welded to an intermediate ribbon which in turn is welded to the terminal. This technique results in the lowest possible tap resistances on the order of tenths of ohms.

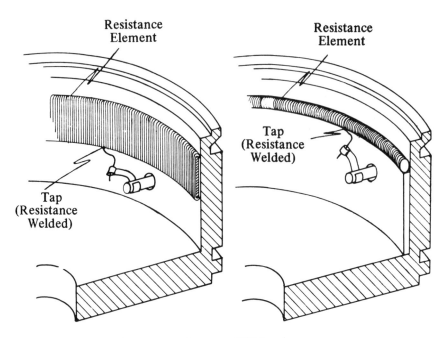

Figure 2–5 Taps *(Courtesy, Bowmar/TIC Inc.)*

Slider Contact and Electrical Takeoff

The slider moves along the contact path of the resistance element to vary the resistance value of either potentiometer arm or to receive the voltage at a selected point along the potentiometer winding. Typical slider contacts are shown in figure 2–6. These include the bar-type contact commonly used with mandrels and the formed-wire contact and button contact commonly employed with wound cards.

As shown in figure 2–7, the slider contact is supported by a rigid arm which, in turn, engages the potentiometer shaft. To operate properly, the slider contact must rotate in a circle that is concentric with the potentiometer winding bonded to the inner surface of the enclosure wall. Pressure of the slider contact should be uniform and carefully adjusted to obtain an optimum compromise between contact noise and wear.

The slider is joined to an external terminal of the potentiometer through a slip-ring circuit such as that shown in figure 2–7. This type of electrical takeoff employs a pair of precious-metal contacts to maintain noise-free contact with a slip ring having an inlaid coin-silver or hard gold surface. Other constructions include mounting the slip ring flat against one of the potentiometer ends, or rotating the slip ring while the contacts are held stationary.

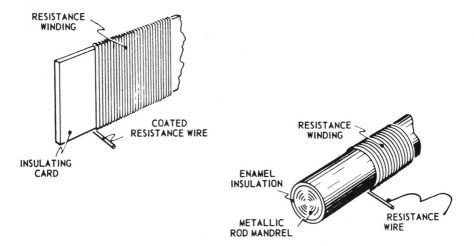

Figure 2-6 Typical Slider Contacts for Mandrel and Card Resistance Elements *(Courtesy, Bowmar/TIC Inc.)*

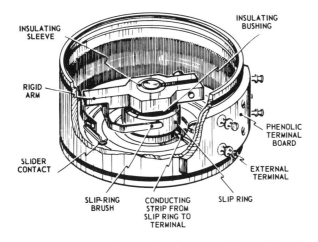

Figure 2-7 Slider and Slip-Ring Assembly Forming Typical Electrical Takeoff for Potentiometer Output *(Courtesy, Bowmar/TIC Inc.)*

External leads are joined to the slider circuit, as well as to the ends of the resistance element and to the taps, through a terminal mounted on the outer surface of the potentiometer. The terminals are rigidly supported in an insulating structure which isolates them from one another and from the potentiometer base. Some advanced potentiometer designs join these terminals to a special plug which permit the convenient plug-in connection of the potentiometer to its electrical circuit.

Shaft and Bearing (See figure 2–8)

Precise rotation of the potentiometer slider is achieved through a shaft, supported in a low-friction bearing. The shaft is commonly of precision-ground stainless steel, and the bearing is usually a precision stainless-steel ball bearing. Occasionally bushings are used to support the shaft; however, the mechanical tolerances specified usually preclude the use of bushings. The semi-precision panel-mounted potentiometer is used where life and accuracy are not critical and usually are of bushing construction.

The upper end of the shaft extends above the face of the potentiometer base and the slider-arm structure is secured to this end. The other end of the shaft extends through the base and is joined to any desired driving mechanism in the system.

Mechanical Stops

In potentiometer applications where full (360°) rotation of the slider is not desired, a stop pin fastened internally to the base is used to limit shaft motion. The pin is generally positioned and dimensioned so that the slider contact re-

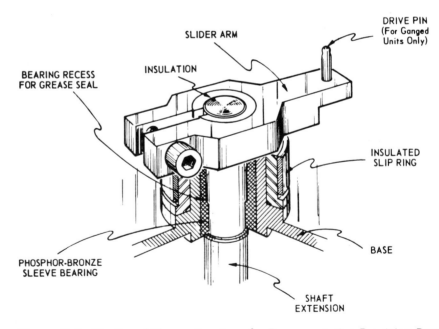

Figure 2–8 Shaft and Sleeve Bearing of a Representative Precision Potentiometer *(Courtesy, Bowmar/TIC Inc.)*

mains on the conducting overtravels at each extremity of slider-arm motion. Dynamic impact imposed by the system should be considered when mechanical stops are required. Inertial loads and system slew speeds may require high-strength stops of 10 to 15 inch-pounds. If the dynamic loads exceed this value, external stops should be installed in the system.

Cover and Enclosure

A sealed cover is used on precision potentiometers, except in ganged potentiometers, in which the bottom of one base fits into the rim of another. Such covering provides an essentially dust-free and moisture-proof enclosure, insuring longer trouble-free life of the internal elements.

The small clearance space between the shaft and through the ball will occasionally allow corrosive elements to enter the potentiometer. This is particularly true if temperature and pressure vary, causing the potentiometer to "breathe" and draw the external atmosphere into the enclosure. Special potentiometers having a shaft that extends out both ends are particularly subject to this type of attack. Special sealing precautions may be taken to diaphragm-seal the shaft and bearing. Other possible points of entry are moisture-sealed.

The use of "O" rings to seal the shaft is usually avoided in low torque applications; "O" rings increase the torque appreciably and also often have "memory," which tends to reposition the shaft away from its set position.

Operating Characteristics

Besides the typical mechanical characteristics such as torque, travel, and contact resistance, each potentiometer has operating characteristics which determine its efficiency and value.

Linearity Unless stated otherwise, the potentiometer is assumed to have independent linearity. It is the maximum deviation from the curve formed by plotting potentiometer electrical output versus mechanical travel and the best straight line that may be drawn through this curve. Factors that determine linearity are: the resolution of the element, the linearity of the element wire, and the stress put on the wire and the placement of the wire when winding the resistance element.

Noise Noise is any electrical signal or disturbance that interferes with the desired electrical output. It may be caused by faulty termination, foreign particles, or oxidation of the element. Equivalent noise resistance expressed in ohms is the effective resistance between the wiper and the resistance element, as the wiper traverses the element. In testing for equivalent noise resistance, a constant current is passed through the wiper circuit. The noise then appears as a random voltage which, when amplified, is visible on an oscilloscope.

Output Smoothness (Non–Wire-wound) Potentiometers   Output smoothness is a measurement of any spurious variation in the electrical output not present in the input. It is expressed as a percentage of the *total applied voltage* and measured for specified travel increments over the *theoretical electrical travel.* Output smoothness includes effects of contact resistance variations, resolution, and other micrononlinearities in the output.

Operating Temperature Range   The operating temperature range is the maximum and minimum ambient temperatures at which a potentiometer may be operated without damage or failure.

Power Rating   This is the maximum that may be dissipated by the potentiometer without permanent damage or excessive temperature rise. It is expressed in watts at a specified temperature in degrees centigrade, with the unit in free air or mounted on a heat sink, as specified. If not specified, it is assumed that a heat sink is not used.

Resolution   Resolution in a wire-wound potentiometer is the average increment of output determined by the number of effective turns of resistance wire on the potentiometer element. It is generally expressed as percent of the reciprocal of the number of turns:

$$\text{Resolution (\%)} = \frac{1}{\text{n (number of turns)}} \times 100$$

Resolution for a film-type or slide-wire potentiometer is theoretically infinite.

Temperature Coefficient   The temperature coefficient is the change in total resistance with a change in temperature of $1°C$. This value may be used to find the change in total resistance throughout any temperature change for which the temperature coefficient remains linear. It is expressed in units $\Omega/\Omega/°C$ and is determined by the following formula:

$$TC(\Omega/\Omega/°C) = \frac{\Delta R}{R \times \Delta t}$$

where TC = the temperature coefficient

R = resistance at initial temperature

ΔR = the change in resistance from initial temperature to final temperature

Δt = the change in temperature from initial temperature to final temperature in degrees C

Voltage Coefficient   This is a measure of the change of resistance with applied voltage. The term is applicable to carbon potentiometers only. The voltage coefficient is expressed in percent per volt and computed from the change in total resistance noted when applying full-rated voltage and one-tenth-rated voltage for a short duration to avoid resistance changes due to self-heating.

Voltage Ratio (Percent)   The voltage ratio is a figure indicating the position of the wiper relative to the terminations of the element expressed in percent of total applied voltage. It is calculated by dividing the output voltage by the total applied voltage times 100.

Potentiometer Types

There are, as you would imagine, a great number of potentiometer types. Several of these are described and illustrated in the next several paragraphs and figures.

Conductive Plastic, Single-Turn (See figure 2-9)   The Bowmar Plastic Pot Model CP20 is representative of a series of precision single-turn conductive plastic servo potentiometers. The conductive track, substrate, and terminations are co-molded into the resistive plastic element. The housing is made of anodized aluminum. The wiper contacts are precious metal. Ball bearings are precision, while the adjustment shaft is made of stainless steel. The adjustment shaft rotates a full 360° clockwise through 350° of element.

Wire-Wound Rotary Trimmer (See figure 2-10)   The Bowmar Rotary Trimmer Model RV 1/2 is a miniature, linear, rotary trimmer potentiometer that has a wide range of resistances (50 to 50K ohms) and relatively high power capability. Its housing is machined aluminum and the adjustment shaft is stainless

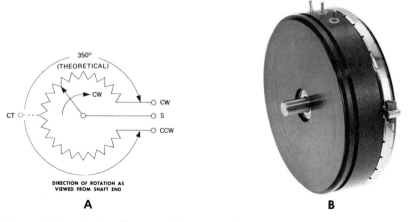

Figure 2-9 A Conductive Plastic Single-Turn Potentiometer *(Courtesy, Bowmar/TIC Inc.)*

Figure 2-10 A Wire-Wound Rotary Trimmer *(Courtesy, Bowmar/TIC Inc.)*

steel. The wiper and contact are made of precious metal. Its shaft bearing is precision-bored, sleeve type. Rotation of the shaft is clockwise through 350°. Terminals are gold-flashed.

Rectilinear Potentiometers (See figure 2-11) The Bowmar Mektron Series PR-1 is a rectilinear potentiometer. The assembly consists of an anodized aluminum housing with a centerless ground stainless-steel shaft. The wiper is made of precious metal and is designed for minimum resistance and noise. The wiper is in contact with both the measuring element and the collector. In the illustration, the shaft line drawing shows the shaft in extended position. Rotation of the shaft moves the wipers linearly through the electrical travel (range) of the potentiometer. The resistance of this potentiometer ranges from 50 to 20,000 ohms per inch at ± 10 percent.

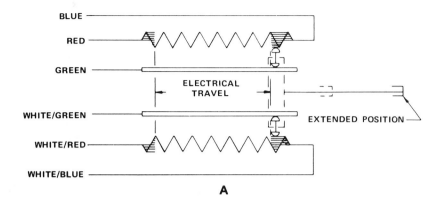

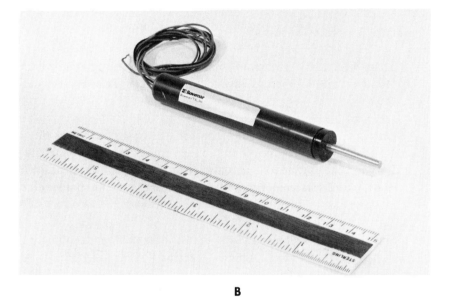

B

Figure 2–11 A Rectilinear Potentiometer *(Courtesy, Bowmar/TIC Inc.)*

SYNCHROS AND RESOLVERS

The term "synchro" is a generic one covering a range of ac electromechanical devices which are used in data-transmission and computing systems. A synchro provides mechanical indication of its shaft position as the result of an electrical input or an electrical output that represents some function of the angular displacement of its shaft. Such components are basically variable transformers. As the rotor of a synchro rotates, it causes a change in synchro voltage outputs.

Operation of the Synchro

Physically, the synchro has the appearance of a motor with a rotor (armature) and a stator (field). The rotor may rotate freely within the stator. There is an electromechanical relationship between the rotor and the stator. The position of the rotor in relation to the stator determines its output. Because of this, the synchro is able to convert rotor shaft position to three-wire electrical signals. If these three wires are connected to three terminals of a second synchro stator, the rotor of the second synchro takes the angle of the first rotor. Thus we have transmitted an angular shaft position from one synchro to a second remote synchro. The applications of this device are many. A typical application is transmission of an aircraft surface-angle position to a cockpit instrument.

Electromechanical Relationships

There are many properties that are peculiar to the individual synchro. However, most properties fall under the heading of electromechanical relationships.

Output Voltage   Output of a control synchro is defined as the open-circuit fundamental line-to-line output voltage at maximum coupling under rated excitation. The output voltage is also given as a voltage gradient in volts per unit of angular displacement. For all types except induction potentiometers, this is expressed as:

$$\text{Voltage gradient} = \text{volts at maximum coupling} \times \text{sine } 1° \text{ (V/°)}$$

Synchro Control Transmitter (CX) (See figure 2–12)   The electromechanical relationships of synchro control transmitters are described as follows:

$$E(S_{13}) = nE(R_{21}) \sin \phi$$
$$E(S_{32}) = nE(R_{21}) \sin (\phi + 120°)$$
$$E(S_{21}) = nE(R_{21}) \sin (\phi + 240°)$$

where $E(S_{13})$ means the voltage between stator terminals S1 and S3. The second subscript (S3, in this case) is usually considered the high terminal and the first subscript (S1), the low terminal. However, the equations remain valid if the first subscript (S1) is considered high and the second subscript (S3) low. Either convention is acceptable as long as consistency is maintained throughout.

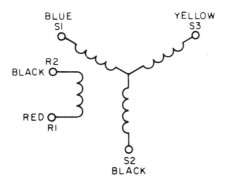

Figure 2-12 Synchro Transmitter Symbols *(Courtesy, The Singer Co., Kearfott Division)*

E(R_{21}) is the voltage between rotor terminals R2 and R1. Other voltages are similarly defined. The letter *n* symbolizes the transformation ratio, while ϕ signifies the electrical angle.

Synchro Control Transformer (CT) (See Figure 2-12) The electromechanical relationships of a synchro control transformer are defined as follows:

$$E(R_{12}) = n\,[E(S_{13})\sin(\phi + 120) + E(S_{32})\sin\phi]$$
$$E(S_{13}) + E(S_{32}) + E(S_{21}) = 0$$

Resolver Transmitters (RX) (See figure 2-13) The electromechanical relationships of resolver transmitters are defined as follows:

$$E(S_{13}) = nE(R_{13})\cos\phi$$
$$E(S_{24}) = nE(R_{13})\sin\phi$$

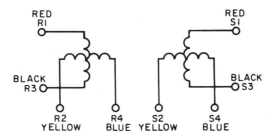

Figure 2-13 Resolver Transmitter Symbols *(Courtesy, The Singer Co., Kearfott Division)*

Resolver Control Transformer (RC)   The electromechanical relationships of a resolver control transformer are defined as follows:

$$E(R_{24}) = n\,[E(S_{24})\cos\phi - E(S_{13})\sin\phi]$$

Electrical Zero   For every rotor position there is a corresponding electrical position. The error at a given rotor position is defined as the mechanical rotor position minus the electrical position.

In general, voltage output from a synchro winding passes through two nulls. In aligning servo systems, it is necessary to zero each synchro physically when the output is zero. Furthermore, in representing error curves graphically, electrical zero is the datum from which all error is plotted. When a pair of synchros is used in a precise application, electrical zero can be selected to provide subtraction of the error curves. By this technique, system error can be made less than the error of the individual units.

On some synchros, the location of electrical zero is denoted by a scribe mark on the shaft which is aligned with an arrow stamped on the housing. For more exact location of electrical zero, a number of procedures are outlined in figures 2–14 through 2–16. A coarse alignment is first necessary to provide the null chosen as EZ.

Impedance (See figure 2–17)   Impedance levels in synchros are chosen for many reasons. Torque synchros usually operate at lower impedance levels than do control synchros because torque output is a direct function of excitation voltage and power. Control transmitters are also designed with lower impedances than their companion control transformers to prevent error voltage feedback when many transformers are loaded on the same transmitters. Usually, in any voltage-generating device, output impedance is held as low as possible to avoid regulation effects.

Any method of determining resistive and reactive elements of impedance can be used. However, proper operating frequencies and voltages should be observed throughout the testing procedure. Impedances usually measured are:

Z_{RO} = rotor impedance with stator OC
Z_{RS} = rotor impedance with stator shorted
Z_{SO} = stator impedance with rotor OC (direct axis impedance)
$Z_{SO} - 90$ = quadrature stator impedance with rotor OC (quadrature axis impedance salient pole synchros)

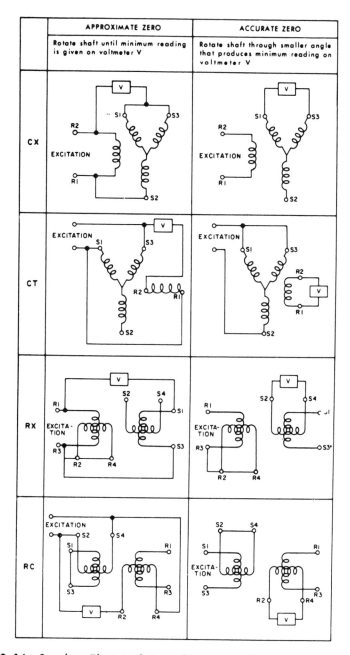

Figure 2-14 Synchro Electrical Zero (*Courtesy, The Singer Co., Kearfott Division*)

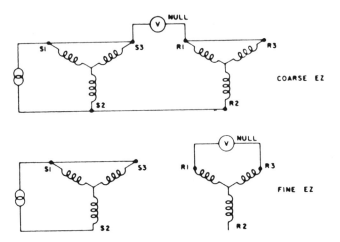

Figure 2-15   Electrical Zero of Control Differential Transmitter and Torque Differential Transmitter *(Courtesy, The Singer Co., Kearfott Division)*

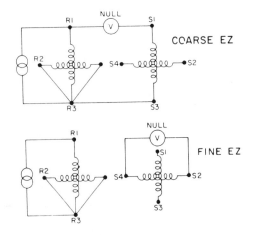

Figure 2-16   Electrical Zero of Resolvers *(Courtesy, The Singer Co., Kearfott Division)*

Impedance levels in resolvers are determined largely by their application. Generally, however, impedances are kept as low as practicable for the physical size and saturation level. Induction potentiometer impedances are governed by many of the same principles that govern resolvers.

Synchro Nulls   In control systems in which the output is used to actuate a servo, the value of the total null voltage is of great importance to the synchro

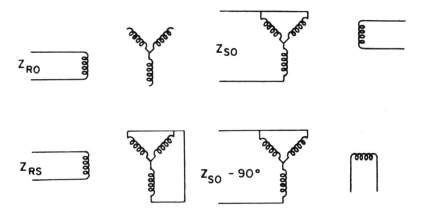

Figure 2-17 Impedance Conventions *(Courtesy, The Singer Co., Kearfott Division)*

user. In a synchro, the total null is the lowest open-circuit residual voltage found as the shaft is rotated about its electrical zero position. It is made up of two basic components—*quadrature fundamental null* and *harmonic voltages.*

Fundamental null is residual voltage having the same frequency as the excitation. This voltage is always in time-phase quadrature with the output voltage at maximum coupling.

Higher time-phase harmonic voltages are the other components of null. They are predominantly third harmonic—i.e., three times the excitation frequency—and are usually combined with the fundamental to yield total average null.

The amount of null that can be tolerated in any servo system varies from system to system, and depends on an amplifier's gain and its ability to discriminate between quadrature null and the in-phase signal, together with its ability to attenuate the higher harmonics. A rule of thumb frequently used in specifying null levels for synchros is: The fundamental quadrature null of the synchro shall be less than one-half the permissible angular error as translated into voltage.

For example, a synchro specified as a 6-minute or 0.1° maximum error unit and having a voltage gradient of 1 volt per degree would normally be permitted to have a quadrature fundamental null of 50 millivolts. Maximum fundamental null = (0.5) (0.1°) (1 volt/deg) = 50 mV.

Total null is normally permitted to be 50 percent to 60 percent greater than the maximum permissible fundamental null. This would amount to 75 mv to 80 mv in the example given above because of reasonable discrimination and attenuation in typical amplifiers.

Torque-Type Synchros

Torque-type synchros are commonly designated as *transmitters* (TX), *differentials* (TDX), and *receivers* (TR), and comprise the simplest system for driving dials and pointers directly without further amplification. Overall accuracy of such systems is limited by friction of the receiver bearing and dial, so emphasis has always been placed upon developing a maximum-output torque to overcome the restraining forces. Like any self-balancing system, dynamic stability must also be considered in a design application.

When the transmission of relatively low torques to a remote location is required, torque synchros are employed. The amount of torque delivered by these simple transducers is sufficient to drive pointers or other light loads with accuracies on the order of one degree.

The elements of such torque systems can be designed in series or parallel chains, as shown in figure 2–18. Though chosen for their small size and mechanical simplicity, such systems are dynamically complex. Stability and steady-state accuracy should be as carefully examined as they are in other closed loops.

Receivers and transmitters are characterized by a three-winding wye-connected stator. That is, each stator winding is spaced—phased 120 degrees from the others—and the external stator terminals are designated as S1, S2, and S3. The rotor consists of a single winding wound upon a salient pole structure, the cross-section of which resembles a pinched circle. Rotor geometry is noted here because it is discernible at the terminals. If stator impedance remains in-

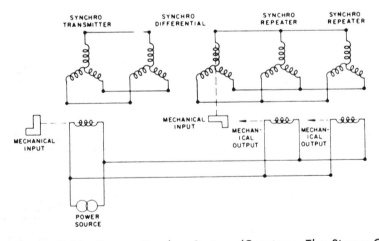

Figure 2-18 Torque Synchro System *(Courtesy, The Singer Co., Kearfott Division)*

dependent of rotor angle, the rotor is round; if it varies between a maximum and minimum value, the rotor is salient pole. In contrast, both the rotor and stator of a differential are three-winding wye-connected, and the rotor cross-section is round.

Energy converted into output torque originates from an alternating power source connected to all transmitters and receivers in the system.

Dynamic Accuracy for torque synchros is defined as the angle by which a repeater shaft lags that of the driving low-error (on the order of 3 minutes) transmitter when the transmitter is rotated at a speed of approximately 2 rpm. Accuracy obtained is on the order of one degree.

Static Accuracy is established by comparing the repeater rotor-shaft position with that of a low-error driving transmitter. The receiver error is read at 15° intervals while the transmitter is stopped and after vibration has been applied to the receiver. Vibration in effect minimizes errors caused by bearing and slip-ring friction and leaves only an electrical error.

Control-Type Synchros

Where large torque and high accuracies are required, control-type synchros are used. In these, a voltage is transmitted for conversion to torque through an amplifier and a servo motor.

Elements of this type of system include:

1. *Control Transmitter* (CX)—A unit in which the rotor is mechanically positioned to transmit electrical information corresponding to angular position of the rotor with respect to the stator. The control transmitter is constructed primarily to operate with control transformers or control differentials.

2. *Control Transformer* (CT)—A unit in which the stator is supplied with electrical angular information and the rotor output is a voltage proportional to the sine of the difference between the electrical input angle and the control transformer rotor angle. The control transformer is constructed primarily to operate with control transmitters or control-differential transmitters.

3. *Control Differential Transmitter* (CDX)—A unit in which the rotor is mechanically positioned to modify electrical angular information received from a transmitter, or to retransmit electrical information corresponding to the sum or difference, depending on the system wiring. The control-differential transmitter is constructed primarily to operate with control transformers.

Control transmitters are often made identical to torque transmitters so that both control transformers and torque receivers can be operated from one transmitter. They are characterized by a salient-pole, rotor-lamination stack having a single-phase winding. The rotor contains two slip rings which conduct excitation to the winding. The stator is a three-phase wye-connected winding.

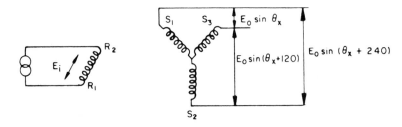

Figure 2-19   Synchro Transmitter Output Function *(Courtesy, The Singer Co., Kearfott Division)*

Figure 2-19 shows the output function of a transmitter which varies sinusoidally with shaft position.

Control transformers use a drum-wound rotor having a distributed single-phase winding. Like transmitter and receiver rotors, these elements also have two slip rings and brushes to conduct power from the control transformer.

Stators in these units are similar in construction to those of transmitters or receivers, but in control transformers the stator is the excitation member or primary. Figure 2-20 shows the addition of a CT to the CS of figure 2-19.

Control-differential transmitters have drum-wound, distributed three-phase, wye-connected windings on their rotors and distributed, three-phase wye-connected stators.

Though not absolutely necessary, it is conventional for the stator to be the excitation member and the rotor to be the output member. Rotors in these devices carry three slip rings and three sets of brushes to conduct output voltages. Figure 2-21 shows the addition of a differential transmitter to the CX–CT combination of figure 2-20.

Resolvers

Resolvers (R) are precision induction-type devices used extensively for coordinate transformation, resolution into components, and conversion from rectangular to polar coordinates. A resolver is essentially a variable transformer so designed that its coupling coefficient varies as the sine or cosine of its rotor position. Usually there are two windings on the rotor and stator at right angles to each other. The rotor windings may be interconnected or isolated from each other with the result that the rotor may have either three or four slip rings. Depending upon the designer's particular application, either the rotor or the stator may be used as primary. Figures 2-22 and 2-23 show output equations for single- and two-phase inputs.

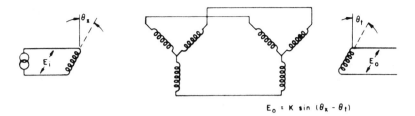

$$E_o = K \sin (\theta_x - \theta_t)$$

Figure 2-20 Synchro CX-CT Combination *(Courtesy, The Singer Co., Kearfott Division)*

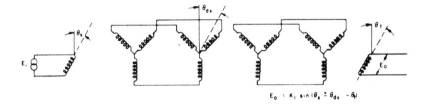

$$E_o = K_1 \sin (\theta_x \pm \theta_{dx} - \theta_t)$$

Figure 2-21 Synchro CX-CT with Added Differential Transmitter *(Courtesy, The Singer Co., Kearfott Division)*

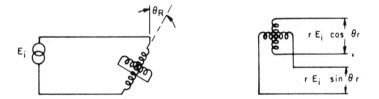

Figure 2-22 Resolver with Single-Phase Input *(Courtesy, The Singer Co., Kearfott Division)*

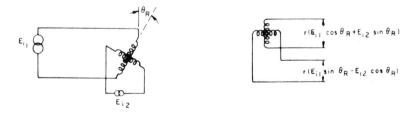

Figure 2-23 Resolver with Two-Phase Input *(Courtesy, The Singer Co., Kearfott Division)*

Induction Potentiometers (See figure 2-24)

An induction potentiometer, or linear synchro transmitter, may be considered a special kind of resolver which provides accurate linear indication of shaft rotation about a reference position in the form of a polarized voltage whose magnitude is proportional to angular displacement, and whose phase relationship indicates direction of shaft rotation.

The principal difference between a resolver and an induction potentiometer is that the latter's output voltage varies directly as an angle and not as the sine of that angle. Its rotation is usually limited to less than $\pm$ 90°.

These devices are analogous to resistance potentiometers, but since they are induction-type components they have less restraining force acting upon their rotors, and hence are capable of providing better resolution. Naturally, induction potentiometers do not require sliders such as those used in resistance types. Therefore, circuit interruptions are eliminated, no wear occurs as a result of rubbing parts, and accuracy is consequently continuously maintained at the original level throughout the operational life of these types of components. These advantageous features are virtually a necessity in certain applications. For example, in gyroscope systems where low restraining torques and low pickoff angular errors are required, wide use is made of these devices. In general, induction potentiometers find use in applications where resistive potentiometers are impractical, principally because of the following features: (1) induction potentiometers, having no wiping contacts, may be used as gyroscope pickoffs since they contribute less spurious friction torque; (2) input and output are isolated; (3) resolution is infinite; (4) noise level is low; and (5) the total angle of travel is limited to less than 180°.

Constructed very much like resolvers, induction potentiometers differ in that their windings and slots are not uniformly distributed, but instead are deliberately modified to produce a linear output. Normally they have only one excitation winding and one output winding. The former is usually carried on the

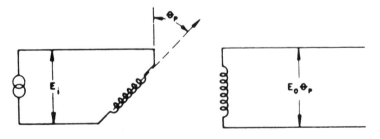

Figure 2-24 Induction Potentiometer Voltage Output *(Courtesy, The Singer Co., Kearfott Division)*

rotor and connected by means of two slip rings and two brushes. Accuracies on the order of 0.1 percent are obtainable.

A Typical Brushless Synchro (See figure 2-25) These units use rotary transformers to couple power into the synchro rotor in place of standard brushes and slip rings. Additional features include the use of extra-wide bearings for increased reliability and improved load-carrying capacity.

This type of design provides system performance advantages in applications where synchros are driven at extremely high speeds or when brush-wiping contact cannot be allowed.

The brushless synchro offers the same electrical characteristics as provided by standard brush-type synchros. The brushless construction is achieved by using spiral hairspring conductors instead of conventional brushes and slip rings.

Since hairsprings require the use of mechanical stops to prevent "overwinding," the synchros are designed with uniquely designed mechanical stops at angles such as 420° for applications requiring a full or more-than-360° excursion, or at angles such as 340° for more limited rotation applications.

The rotary transformer coupled units exhibit increased phase shift and transformation ratio temperature coefficient when compared to standard synchros and resolvers.

A Typical Brushless Resolver (See figure 2-26) The industrial brushless resolvers are designed to be coupled directly to factory equipment, such as a

Figure 2-25 A Typical Brushless Synchro *(Courtesy, The Singer Co., Kearfott Division)*

Figure 2–26 A Typical Brushless Resolver (*Courtesy, The Singer Co., Kearfott Division*)

machine tool lead-screw, for use as the resolver position feedback unit in closed-loop NC machine-tool applications.

Basically, these are brushless single-speed and multispeed resolvers, built using rugged oversized shafts, bearings, and lamination assemblies to provide a durable industrial product.

A rear-sealed connected and double-sealed front bearing are used to protect the unit during operation in a machine-tool/industrial environment.

Since multispeed resolvers exhibit "N" complete electrical cycles (speed) of output signal for each 360° of shaft rotation, they do not require the use of gears to obtain scale variations, as is required in other types of designs. This also results in the elimination of any need for periodic gear backlash adjustments and/or possible related gear, bearing, or internal coupling breakdown problems.

The single-speed and multispeed brushless units use rotary transformers to couple power into the rotor in place of standard brushes and slip rings. Further, the omission of brushes eliminates any possibility of brush bounce that could be interpreted as an error signal by the digital controller.

This type of design provides system performance advantages in applications where the units are driven at extremely high speeds or in an explosive atmosphere when brush-wiping contact cannot be allowed.

Figure 2-27 A Typical Brushless Linear Induction Potentiometer *(Courtesy, The Singer Co., Kearfott Division)*

A Typical Brushless Linear Induction Potentiometer (See figure 2-27) This unit provides a dc output signal proportional to shaft displacement (i.e., $V_O = K_\theta$) over the rotational range of $\pm 60°$. It combines a brushless linear induction potentiometer and a hybrid thick-film carrier-signal conditioning electronic module to provide a dc sensor with a high scale factor, high linearity, and excellent reliability. Because this unit is brushless, it provides an increased life and eliminates the noise problems associated with brush-type dc potentiometers.

THE LINEAR VARIABLE DIFFERENTIAL TRANSFORMER (LVDT)

The linear variable differential transformer (LVDT) is the primary mutual inductance element. The LVDT produces an electrical signal that is proportional to the linear displacement of a movable armature or core.

The LVDT has a simple construction. Basically there are two elements involved with the LVDT, the armature and the transformer. See figure 2-28A. The transformer has a stationary coil enclosed in a protective magnetic shield. The armature then moves within the hollow core of the coil.

The coil has a primary winding in the middle and two secondaries, wired in series opposition. (See figure 2-28B.) When the primary is energized by an ac

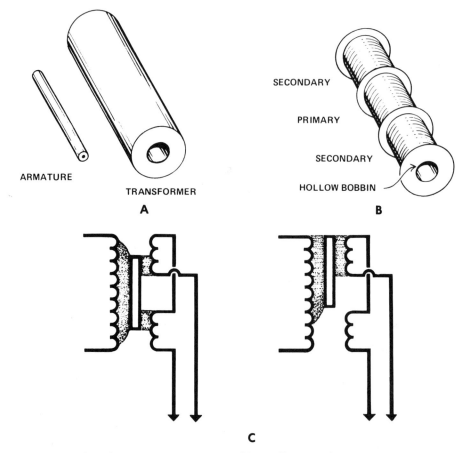

SECONDARY

PRIMARY

SECONDARY

ARMATURE

TRANSFORMER

HOLLOW BOBBIN

A

B

C

Figure 2-28 The Basic Linear Variable Differential Transformer (LVDT) *(Courtesy, Automatic Timing and Controls Co.)*

current, the armature—made of a closely controlled magnetic material—induces a voltage from the primary to the secondary windings. The position of the armature within the core of the coil determines the level of the voltage at each secondary: if the armature is placed precisely midway between the two secondaries (null position), the induced voltage in each secondary is equal and opposite, and there is no output. (See figure 2–28C.) As the armature is moved in either direction away from null figure 2–28D, the LVDT produces an output voltage that is proportional to the displacement of the armature from null and whose phase relationship with the primary supply shows whether the armature has moved nearer one end or the other of the coil. Thus, for each position of the armature, there is a definite output voltage, different in level and polarity than for any other position, no matter how slight the difference.

The schematics in the figure are representative of the ac/dc LVDT. The ac/dc LVDT has a demodulator in the output legs of the secondary. The demodulator is solid state and appearance of the LVDT does not change because of the small size of the demodulator. The size of the output signal shows how far the armature is from null. The positive or negative polarity of the output shows whether the armature is nearer one end of the LVDT or the other.

Special LVDT Characteristics Because of its special characteristics, the LVDT has distinct advantages over other devices used for motion mechanization of transducers. Some of these advantages are as follows:

1. There is no friction or hysteresis, since there is no mechanical contact between windings and armature.
2. There is no mechanical wear, hence virtually infinite life.
3. Linear output assures accurate measurement with direct readout instruments.
4. There is infinite resolution, limited only by the readout and control equipment.
5. There is complete electrical isolation of output from input, permitting addition or subtraction of signals without buffer amplifiers.
6. High-level output simplifies circuitry.
7. Over-ranging does not cause any damage or permanent change in characteristics.
8. The LVDT is rugged and shock-resistant, and virtually free of maintenance.

LVDT Linearity Absolute linearity is defined as the largest deviation of transducer output from the theoretical ideal, measured between zero and full-scale displacement and expressed as a percent of the full-scale output.

LVDT Sensitivity At high-frequency excitation, LVDT sensitivity is practically independent of changes in ambient temperature. Sensitivity ratings are therefore measured at constant voltage, and expressed as millivolts output per volt of excitation per 0.001 inch of displacement (mV/V/0.001 inch).

At low-frequency (60Hz) excitation, the voltage sensitivity of LVDTs changes with temperature. It is therefore normally expressed at constant current (V/A/inch), which may be converted to mV/V/.001 inch as follows,

$$\text{mV/V/.001 in.} = \frac{\text{V/A/in.}}{\text{primary impedance}}$$

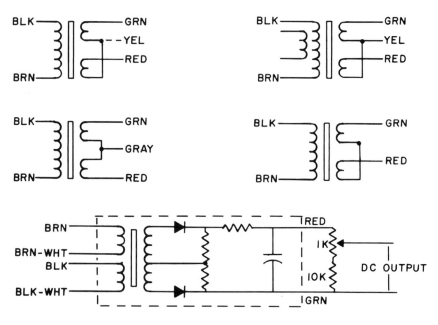

Figure 2-29 LVDT Coil-Winding Configurations *(Courtesy, Automatic Timing and Controls Co.)*

LVDT Coil-Winding Configurations   The LVDT can be wound in many ways to accommodate special applications. Some of these windings are illustrated in figure 2–29.

A Typical Linear Variable Differential Transformer (See figure 2–30)   The Automatic Timing and Controls Co. (ATC) Series 6020 LVDT is typical of the linear-position transducers on the market today.

This displacement transducer uses a differential transformer to translate a linear mechanical displacement, such as lever position, roller position, mechanical feed, cam position, etc., into an exactly proportional electric signal for local or remote transmission and evaluation. The armature of the transformer is mounted on the spring-loaded plunger that is placed against the device to be measured. Thus, any movement of the transducer tip and armature provides a direct proportional electrical signal for indication, recording, or control.

The transducer consists of a hardened adjustable contact tip, in a locking assembly, mounted on a spring-loaded shaft. The shaft travels through and is held in a linear position by an oilite bearing. Attached to the inside end of the shaft is the transformer armature. The transformer coil is held in position by set

Figure 2-30   A Typical LVDT *(Courtesy, Automatic Timing and Controls Co.)*

screws through the transducer housing. The transducer cylindrical housing is cadmium-plated.

The differential transformer is a cylindrical-shaped coil with a hollow center, through which an armature, or piece of magnetic metal, is moved without contacting the coil to produce an output voltage proportional to the armature position. Having no linkages or friction, the differential transformer gives practically infinite life and resolution for measuring straight-line displacement.

REFERENCES

Reference data and illustrations in Chapter 2 were supplied by state-of-the-art manufacturers. Permission to reprint was given by the following companies:

Potentiometers—Bowmar/TIC Inc., Newbury Park, California
Synchros and resolvers—The Singer Co., Kearfott Div., Little Falls, New Jersey
LVDTs—Automatic Timing and Controls Co., King of Prussia, Pennsylvania
All copyrights © are reserved.

3

Force Transducers: Strain Gages, Load Cells, and Weigh Cells

STRAIN GAGES

The purpose of a strain gage is to detect the amount or length displaced by a force member. The strain gage produces a change in resistance that is proportional to this variation in length. A basic strain-gage transducer function was illustrated in the figure 1–19. You may wish to refer once again to that figure.

The strain gages are usually installed as part of a Wheatstone bridge for electrical-circuit applications.

There are two basic strain-gage types. These are the bonded and the unbonded. The *bonded gage* is entirely attached to the force member by an adhesive of some sort. As the force member stretches in length, the strain gage also lengthens.

The *unbonded gage* has one end of its strain wire attached to the force member and the other end attached to a force collector. As the force member stretches, the strain wire also changes in length. Each motion of length, either with bonded or unbonded gages, causes a change in resistance. Strain gages are made from metal and semiconductor materials. Strain gages are accurate, may be excited by alternating or direct current, and have excellent static and dynamic response. The signal out of a strain gage is very small, but this disadvantage may be corrected with good periphery equipment.

Unbonded Strain Gages (See figure 3–1)

The sensor of an unbonded gage consists of wire, usually 0.3 mil to 0.5 mil in diameter, wrapped around nonconductive sapphire posts. In some designs, the

Semiconductor Strain Gages (See figure 3-3)

The recent technologies developed for manufacture of transistors and integrated circuits have now been applied to sensors for pressure transducers. IC manufacturing techniques can be employed to build, calibrate, and test pressure transducers. High-accuracy, stable-pressure transducers can be produced in high volume at very low costs.

Within the semiconductor or piezoresistive family of strain gages, two types of sensors are employed: the bar gage and the diffused gage. The *bar-gage* transducer is nearly identical to the bonded-foil strain-gage instrument. Its gages are individually bonded with epoxy to a mechanical structure, such as a beam or diaphragm. The *diffused-gage* transducer uses a silicon element for the mechanical structure and the strain gage is an integral part of the silicon element rather than being bonded to it—i.e., the gage is diffused into the structure.

In the bar gage, an impurity such as boron is introduced into the silicon crystal as it is grown. In a bar gage, the entire piece of silicon is an active gage. The individual gages are then electrically matched, as nearly as possible; then four legs of a bridge circuit are cemented to the pressure sensor (either a diaphragm or bending beam). Although the four-gage elements made from silicon-based semiconductors exhibit high gage-factor levels (20 mV/V), they suffer inherent limitations associated with other bonded strain gages—primarily zero shift and creep from elasticity and different coefficients of the bonding cement and sensor materials.

Although the bar gages are somewhat easier to produce, considerable assembly labor and element matching is required to produce a good transducer (figure 3-3A.)

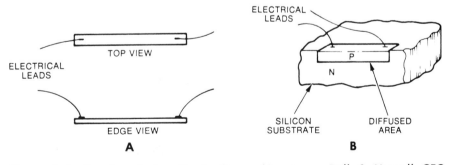

Figure 3-3 Semiconductor Strain Gage (*Courtesy, Bell & Howell CEC Division*)

The diffused gage employs a multi-setup process of cleaning and polishing of the sensor surface, and the application of photoresist, gage deposit, masking, electrical terminal depositing, and lead attachment (figure 3–3B).

Automation of the gage manufacturing process provides consistent results at relatively low cost. If the pressure diaphragm and sensor are properly designed, excellent performance characteristics are obtainable. Pressure-sensor design considerations appear to be a limitation with some of the transducers offered. Successful integration of good pressure-sensor designs and semiconductor practices will result in very stable quality instruments.

Strain-Gage Compensation (See figure 3–4)

Strain or shear is induced by the pressure sensor upon the wire or crystalline strain-gage element. Each active element exhibits a resistance change, which is additive in effect, within a Wheatstone bridge circuit.

The resistance change, compared to the unstrained element resistance, is often referred to as gage factor:

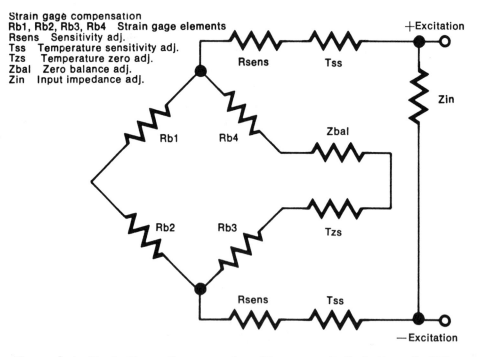

Strain gage compensation
Rb1, Rb2, Rb3, Rb4 Strain gage elements
Rsens Sensitivity adj.
Tss Temperature sensitivity adj.
Tzs Temperature zero adj.
Zbal Zero balance adj.
Zin Input impedance adj.

Figure 3–4 Strain-Gage Compensation *(Courtesy, Bell & Howell CEC Division)*

$$\text{Gage factor is expressed as } GF = \frac{\Delta R / R}{\Delta L / L}$$

Where GF is gage factor

ΔR = change in resistance

R = unstrained element resistance

ΔL = change in element length

L = unstrained element length

Different types of gages exhibit different gage factors. Gage factor is important in the proper design of transducers. High-amplitude signal changes are desirable, provided other performance characteristics such as temperature sensitivity are acceptable to the measurement.

Gage factors of the various types of strain gages can be compared:

Type of Sensor Gage	Gage Factor
Unbonded wire	4
Bonded foil	2
Thin film	2
Diffused semiconductor	80–150
Bonded-bar semiconductor	80–150

From the user's point of view, transducer output is the criterion. Gage factor is related to output level and sensitivity, but variances from unit to unit are adjusted to a nominal output sensitivity with resistors placed in series with the gages.

Figure 3-4 shows compensation resistors and the circuitry location in which they are employed. Manufacturers can adjust output for:

1. Change in zero-output level
2. Full-scale level (sensitivity or span)
3. Temperature effects on zero
4. Temperature effects on full scale (sensitivity or span)
5. Input/output impedance matching

THE LOAD CELL

A load cell is a transducer device that produces an output that is proportional to an applied force. The load cell facilitates the accurate measurement of force and weight. Some load cells are strain-gage types. This section describes the load cell and its loading and weighing function.

Description of an LVDT Load Cell

There are, of course, several types of load cells. The concepts of all types are similar. A typical load cell is the variable differential transformer (LVDT) load cells. The unit is based on a cantilever beam principle (see figure 3–5).

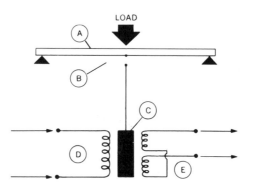

Figure 3–5 Operation of an LVDT Load Cell *(Courtesy, Automatic Timing and Controls Co.)*

The heart of the LVDT load cell is the deflection element, which is a series-parallel arrangement of multiple center-loaded beams. When an axial load is applied to a center-loaded beam (A) in figure 3–5, it produces a proportional linear deflection (B) that is sensed by the armature (C) of a differential transformer (LVDT). As the armature moves relative to the primary coils (D) of the LVDT, it changes the distribution of the magnetic field and consequently the voltage that is induced into the secondary coils (E). Thus, for every change in applied load there is a specific deflection and output voltage of the load cell.

An LVDT Load Cell (See figure 3–6)

The figure illustrates an Automatic Timing and Control Co. (ATC) Series 6004 load cell along with its LVDT element.

The LVDT load cell is a practical unit with a high electrical output. This fact greatly simplifies its instrumentation requirements. The same task, if accomplished by a strain gage, may take sophisticated amplification, which would result in higher instrumentation costs.

Several of these load cells may be placed in electrical series. In contrast, multiple-strain-gage installations require isolated dc power supplies for each unit.

Figure 3-6　An LVDT Load Cell *(Courtesy, Automatic Timing and Controls Co.)*

The load cell is used to weigh high capacities. However, the load cell itself does not contain such things as flexures, dashpots, and mounting platforms that are required for a weighing machine. The load cell, as was previously stated, is a precise spring and LVDT that generates a signal that is proportional to the force applied to the cell. The electrical element (the LVDT) conveniently produces its output signal in isolation to the power source.

The ATC Series 6004 is ideally suited to filling machines, weighing conveyor machines, and torque and force measurements.

LVDT Load-Cell Applications (See figure 3-7)

The illustration presents four typical applications of the LVDT load cell. Best results are obtained when the load is applied in tension through a flexible coupling. Side motion should be held to a minimum (± 0.005 inch). Figure 3-7A illustrates the load cell in application with a filler or hopper for remote indication

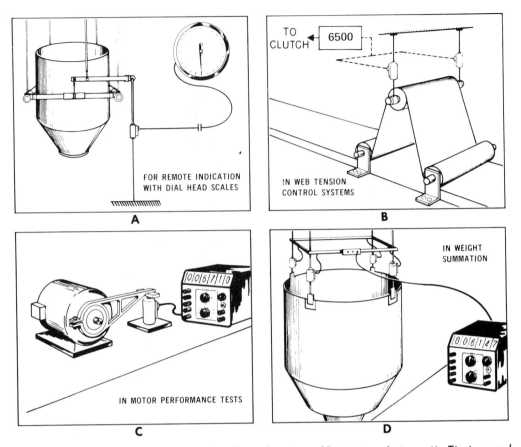

Figure 3-7   LVDT Load-Cell Applications *(Courtesy, Automatic Timing and Controls Co.)*

with dial-head scales. Figure 3-7B shows the load cell used in conjunction with weight module to monitor and control web tension.

In figure 3-7C, the load cell is used in a motor performance test along with a dashpot. Finally, in figure 3-7D, the LVDT load cell, combined with a dashpot, performs weight-summation functions. The dashpot is practical to dampen frequencies that may otherwise cause difficulties.

Description of a Strain-Gage Load Cell

The standard strain-gage load cell in industry has been the column type. In order to describe the low-profile load cell, let us compare it with a column-type load cell (see figure 3-8).

The column strain-gage load cell is shown in cutaway in figure 3-8A. The

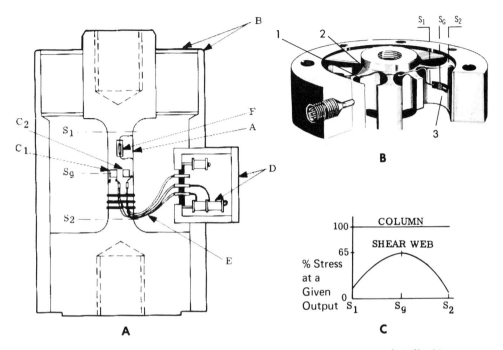

Figure 3–8 Low-Profile Versus Column Strain-Gage Load Cells *(Courtesy, Interface Inc.)*

low-profile shear-web load cell is illustrated in cutaway in figure 3–8B. Figure 3–8C is a graph comparing stress percentages.

Column load cells have two arms, of a four-arm bridge, aligned parallel to the load axis, and two arms aligned at 90° (C_1 and C_2) to measure poisson strain. Poisson strain is the ratio of side to lengthwise strain. The four-arm bridge is 2.6 times the output of a single arm aligned with the principal axis.

The shear-web design has principal strains of equal magnitude and opposite sign, permitting output of the four-arm bridge (3) to be four times the output of a single arm aligned with the principal stress. This greater efficiency of the shear web (153 percent of column type), results in either a higher output for strain level or lower strain level for the same output.

The column load cell is typically at heights of 6 inches to 24 inches. The low profile is 1 inch to 3½ inches in height.

Bending, torsion, and side loads produce high fatigue life. The column-type load cell (A) has low resistance to bending moments and side loads, necessitating a supporting outer shell and diaphragm system (B). The low-profile shear cell has inherent resistance to extraneous forces, which permits less stringent mounting alignment and reduces the possibility of errors.

The cross-section area of a column changes with the load and therefore is nonlinear. Compensation for this nonlinearity is made with a semiconductor strain-sensitive resistor (F). The shear-web design provides equal resistance change in all arms of its bridge, eliminating the need for a compensation resistor.

In column load cells, the strain level (S1 to S2) is constant at maximum strain (see figure 3–8C). In shear-web designs, the webs are contoured to produce peak strain only under gage grids (center stress).

Column cells get longer in tension and shorter in compression; also, the cross-sectional area decreases in tension and increases in compression. This produces higher bridge sensitivity in tension than in compression. In comparison, the shear web is identically deformed in tension and compression.

As a conclusion, strain-gage load cell performance has several necessary criteria:

1. The cell should be as compact as possible to fit into tight places.
2. The load cell should resist extraneous forces in order to detect those forces that are being measured.
3. A high output is desirable as in any transducer.
4. The unit should be stable in terms of both temperature and barometric pressure.
5. Outputs should be symmetrical.
6. A final criteria is long life (as with any transducer device).

A Strain-Gage Load Cell

Figure 3–9 illustrates an Interface Inc. Model 1210 shear-web strain-gage load cell.

This particular unit is used for weighing, fatigue testing, and/or measuring thrust, torque, and force. These uses are, of course, similar to other load cells such as the LVDT.

There may be as many as four load readings taken from each transducer. These include readout and control, multiple control, statistical averaging, and redundancy. Each load cell has as many as four electrically isolated bridges.

The shear-web load cells may also simultaneously measure thrust (T) and moment (M), as shown in figure 3–9. These measurements are used for single-point center-of-gravity determination and force-vector studies. Thrust is measured parallel to the central axis. Moments are measured about two perpendicular axes intersecting at the center of the transducer and perpendicular to the thrust axis.

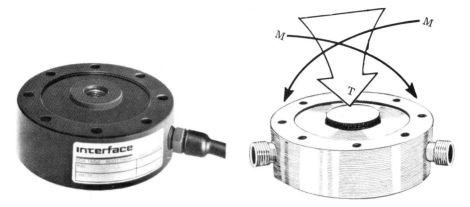

Figure 3-9   A Shear-Web Strain-Gage Load Cell *(Courtesy, Interface Inc.)*

Strain-Gage Load-Cell Error (See figure 3-10)

Load-cell errors are usually controlled and known so that compensation for error may take place. Manufacturers perform tests to determine precisely what the error may be. Testing is proprietary and required by company directives. Three of these tests are illustrated in figure 3-10.

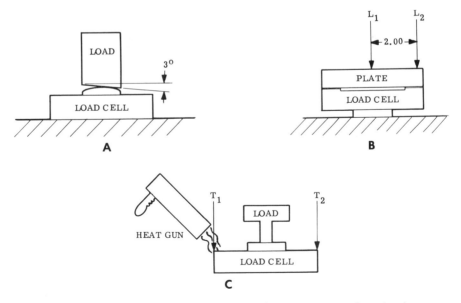

Figure 3-10   Strain-Gage Load-Cell Error *(Courtesy, Interface Inc.)*

The illustrated test (figure 3–10A) is typical of a load applied by a weight bridge caused by the bending of the beams. This type of load applies a bending moment, and it moves the load point off-center. In the particular test illustrated, the error was less than 0.03 percent.

The test in figure 3–10B demonstrates "off-center loading," which is often encountered, especially in fatigue testing, when one side of the specimen fails. This specific test is conducted using a load of 50 percent rated range to keep the bending moment to 100 percent of rated range when loading at L2. The same load applied at point L1 and at any point on a circle scribed by point L2 caused errors less than 0.1 percent.

The test in figure 3–10C demonstrates the load cell's resistance to thermal gradients. A load cell is subjected to heat at point T1 until there is a 100 percent difference between points T1 and T2. The maximum error recorded during this test was 0.1 percent.

Strain-Gage Load-Cell Applications (See figure 3–11)

Strain-gage load cells are used in weigh-scale pits, conveyor scales, batching plants, and other such force and weight applications. They are rugged enough

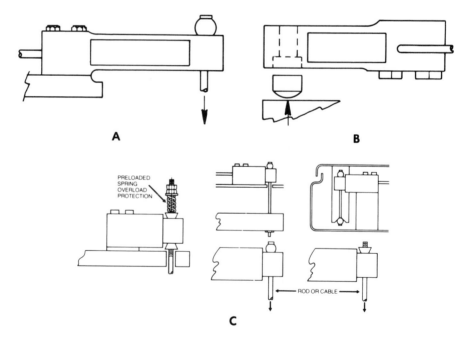

Figure 3–11   Strain-Gage Load-Cell Installations *(Courtesy, Interface, Inc.)*

to withstand all weather or washdown situations. In general, these strain-gage load cells can be used anywhere a readout instrument is used. The force to be measured is applied to the active end of the cell. This is to eliminate possible errors caused by cable interaction. In figure 3–11A and 3–11B, large-capacity (50-to 10,000-pound) strain-gage load-cell installations are described. In figure 3–11A, the cell is connected to a support structure by bolts and a pulling force is applied to the active side of the cell. In figure 3–11B, the cell is again attached to a support structure by bolts and a pushing force is applied to the active side of the cell.

In figure 3–11C, small-capacity (5 to 250 pounds) load-cell installations are illustrated. The reader will note that overload protection can be supplied with the use of a preload spring.

Load cells are mounted in accordance with the manufacturer's recommendations. A fundamental good practice is that there be one and only one load path. This load path must be through the load axis of the load cell.

WEIGH CELLS

A weigh cell is a transducer used in filling-by-weight systems, in batch-weighing systems, and in check-weighting systems. The weighing system may use several types of transducers, the most prominent of which is the linear variable differential transducer (LVDT). In the next several paragraphs we shall discuss the LVDT-type weigh cell and its use with weighing systems.

Description of an LVDT Weigh Cell (See figure 3–12)

A unique weigh-cell device is one that uses the LVDT as a transducer to monitor high-speed fill operations.

The weigh cell is a relatively simple device. It has no knife edges or bearings. It consists of five primary components: a stationary mounting plate, flexure plates, a precision spring, an LVDT, and an oil-filled dashpot.

Let's discuss these parts in a cursory manner. The stationary mounting consists of a cast housing/frame and a bracket. The cast housing is rigid. It firmly holds the fixed ends of the flexures and mount for the LVDT coil and the top of the range spring. The bracket is firmly attached to the housing. The bracket holds the top of the range spring and the stationary coil for the LVDT. The flexures are located on the top and the bottom of the unit. The flexures are fastened at one end to the housing and hold the yoke and the LVDT coil in position while allowing movement up and down.

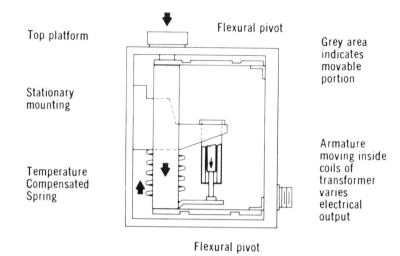

Top platform

Flexural pivot

Grey area
indicates
movable
portion

Stationary
mounting

Armature
moving inside
coils of
transformer
varies
electrical
output

Temperature
Compensated
Spring

Flexural pivot

Figure 3-12   Description of an LVDT Weigh Cell *(Courtesy, Automatic Timing and Controls Co.)*

The top platform is the mounting device for the weigh cell. The top platform attaches to a mount plate, a platform, a hopper, etc. The platform moves downward when weight is placed on it. A flexible boot seals the cell at the top platform so that it can be washed down. The seal allows free movement of the platform. A yoke attached to the platform moves down when weight is placed on the platform and moves up again when the weight is removed.

The precision spring exerts upward force against the weight on the platform to reposition the yoke when weight is removed.

The coil of the LVDT is firmly held in place by the bracket. Its armature, inside the coil, is free to move up and down with the yoke when weight is placed on the platform.

Now let's see how the entire operation takes place. When weight is placed on the platform, the platform, yoke, and armature move downward a few thousandths of an inch against the upward thrust of the big coil spring.

When 2 ounces are placed on a 4-ounce weigh cell (50 percent of its live range), its platform will move down 0.030 inch; and when 16 pounds is placed on a 32-pound weigh cell (50 percent of its live range), its platform will move down 0.030 inch.

Since the armature's position within the coil is directly related to this downward movement of the platform, the LVDT's output signal is proportional to the weight on the platform, except when some preload has been adjusted onto the cell.

Figure 3-13 An LVDT Load Cell *(Courtesy, Automatic Timing and Controls Co.)*

A typical weigh cell is the Automatic Timing and Controls Company weigh cell illustrated in figure 3–13.

Weigh-Cell Applications (See figure 3–14)

Two weigh-cell applications are illustrated in the figure. Figure 3–14A is a block diagram of a weigh cell used with a fill-by-weight system. This simple system works well when all containers weigh the same and when the characteristics of the filler can provide the desired accuracy with a single final cutoff point. When the empty container is placed on the weigh cell, a switch with a holding circuit wired through the relay module starts the filler. The filler cuts off when the preset amount is in the container. The set point can be adjusted up or down to obtain a desired new amount in the container.

Figure 3–14B is a check-weighing application. Separate relay modules, each with a set-point potentiometer, provide individual setting of underweight reject and overweight reject points. When different batches of different weights are handled, both of the set potentiometers must be readjusted to establish the new reject points.

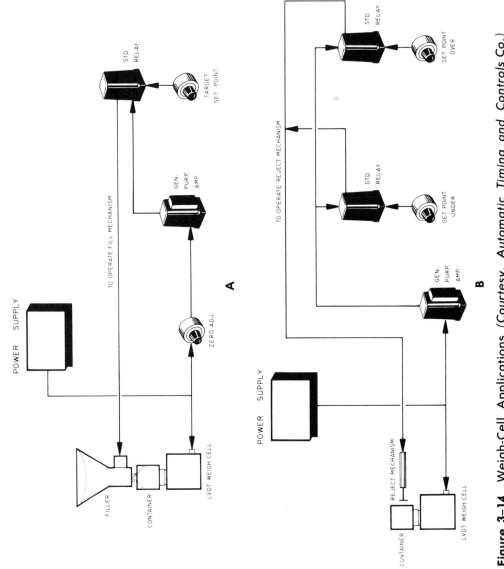

Figure 3-14 Weigh-Cell Applications (*Courtesy, Automatic Timing and Controls Co.*)

90

REFERENCES

Reference data and illustrations in Chapter 3 were supplied by state-of-the-art manufacturers. Permission to reprint was given by the following companies:

Strain gages—Bell and Howell, CEC Division, Pasadena, California
LVDT load cells and weigh cells—Automatic Timing and Controls Co. (ATC), King of Prussia, Pennsylvania
Strain-gage load cells—Interface, Inc., Scottsdale, Arizona
All copyrights © are reserved.

4

Vibration: Acceleration, Displacement, and Velocity

One of the major problems encountered with machines in motion or machines that have rotating members is vibration. Vibration (also called dynamic motion) is clearly singled out to be the most monitored of all the unwanted machine parameters.

Testing is periodically accomplished on machines to analyze structural test points and calibration stability. Machines are constantly monitored for loose screws, broken components, and faulty connections. Indeed, vibration is a serious problem, one which has become a major study of many transducer manufacturers.

The purpose of this chapter is to provide some background basics for transducers, sensors, and detectors that are involved with vibration monitoring. We shall look at these parameters individually and collectively, along with other parameters that are involved. Although this is not a physics book, it seems apropos to the subject.

VIBRATION TERMS

There are essentially three major parameters involved in vibration analysis. These are acceleration, displacement, and velocity. *Acceleration* is monitored in terms of peak g's, *displacement* in terms of peak to peak mils, and *velocity* in peak inches per second. Each one of these parameters also has a time base and a sinusoidal relation involving frequency of vibration. Monitoring of dynamic motion usually involves test equipment whose outputs are in display form, such as oscilloscopes. Monitoring forms are frequency, amplitude, phase angle, and the form and shape of the displayed waveform. Each of the parameters of vibration has exact mathematical relationships which we shall attempt to unravel here in this chapter.

Vibration

Vibration is the basic term that defines the act of dynamic motion. Webster defines vibration as the periodic to and fro motion or oscillation from rest or static position. While vibration may be purposefully caused, such as with a collating machine, it is usually unwanted. Vibration can be detected even in small quantities with the use of precision transducers placed in selected positions on a machine or other apparatus.

Most preventive maintenance programs involve the periodic checking of machines or other apparatus to ensure that they do not vibrate beyond allowable limits. Small vibrations can grow if not checked at their fault source. Most large machines have transducers permanently installed to constantly monitor vibrations. In this manner, the machine fault may be corrected before it causes shutdown of equipment and standby of personnel until repair is accomplished. Transducer manufacturers design their product so as to fit in tight places, inaccessible locations, unusual vibration zones, and so as to detect even the smallest vibration out of the normal.

Vibration occurs in three forms: These are displacement, acceleration, and velocity. Vibration is measured in terms of *frequency, amplitude, phase angle, form,* and *shape.*

Displacement (D)

Displacement is the actual distance that an object moves from equilibrium or rest position. Displacement is measured in inches peak to peak. The relationship of displacement (D) with acceleration (A) and velocity (V) is mathematically explained thusly:

$$\text{Displacement (D)} = 1957 \frac{A}{f^2} \text{ inches, peak to peak}$$

where f = frequency in hertz

A = acceleration in peak g's

$$\text{Displacement (D)} = 0.318 \frac{V}{f}$$

where f = frequency in hertz

V = velocity in inches/second, peak

Velocity (V)

Velocity is the rate of change of displacement with respect to time. Velocity is measured in inches-per-second peak. Furthermore, velocity is related to displacement (D) and acceleration (A) in the following manner:

Velocity (V) = πfD inches/second, peak

where π = 3.1416 (constant)

f = frequency in hertz

D = displacement in inches, peak to peak

Velocity (V) = $\dfrac{61.44g}{f}$

where   f = frequency in hertz

A = acceleration in peak g's

Acceleration (A) In Relation to Displacement and Velocity

Acceleration is the rate of change of velocity with respect to time. Acceleration is measured in g's, where one g = 32.2 feet/second/second or 386.1 inches/second/second peak. Furthermore, acceleration (A) is related to displacement (D) and velocity (V) in the following manner:

Acceleration (A) = 0.0511 f^2D in g's

where   f = frequency in hertz

D = displacement in inches, peak to peak

Acceleration (A) = 0.0162 Vf

where   f = frequency in hertz

V = velocity in inches/second, peak

VIBRATION ANALYSIS

A Vibration Study

It probably has occurred to most of the readers that machinery should not vibrate—that is, unless it is made to vibrate. Generally speaking, there are methods for the monitoring of the major parameters. These include the amplitude of vibration in peak-to-peak mils displacement, peak inches per second velocity, and peak g's acceleration.

An interesting relationship between the three basic parameters of vibrations was made by Peter C. Sundt, president of Metric Instrument Co., Houston, Texas. Sundt compared a displacement sine-wave plot of a vibrating object with its velocity and acceleration sine-wave plots (see figure 4–1). The vibration displacement point was made to move with simple sinusoidal (harmonic) motion. The frequency was made to double with respect to time. In this fixed case situation, velocity amplitude increased as frequency increased. The velocity peaks were 90° out of phase with the displacement peaks. Velocity peaked at zero displacement. Displacement peaked at zero velocity. Similarly, the acceleration plot increased in amplitude with frequency. Maximum acceleration occurs as the vibrating point passes through its extreme position and when velocity passes through zero. Acceleration is zero at maximum velocity where the vibrating point is passing through its equilibrium point. In other words, the acceleration sine wave lags behind the velocity sine wave by 90° and the displacement sine wave by 180°.

Peak acceleration is proportional to peak displacement times the square of the frequency. If peak displacement is held constant and frequency is doubled, peak acceleration will be quadrupled.

Expressed mathematically, here is the summary of these relationships between displacement, velocity, and acceleration in sinusoidal vibration. Instantaneous displacement in sinusoidal vibration is defined by the following formula:

$$x = X \sin \omega t$$

where x = instantaneous displacement

X = peak displacement

ω = vibration frequency in radians per second ($2\pi f$ radians per second in angular measure where f equals frequency)

t = time in seconds

Instantaneous velocity in sinusoidal vibration is defined by the following formula:

$$v = (\omega X) \cos \omega t$$

where v = instantaneous velocity

ωX = peak velocity

ω = vibration frequency in radians per second ($2\pi f$ radians per second in angular measure when f equals frequency)

t = time in seconds

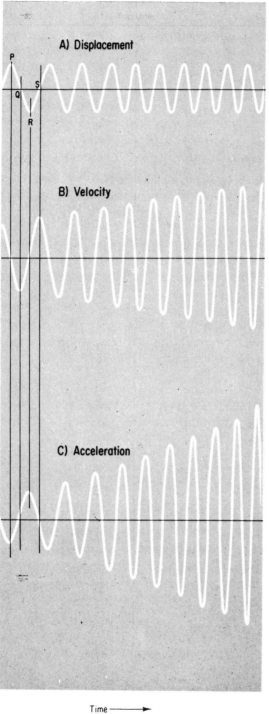

Figure 4-1 Sinusoidal Vibration Plots of Displacement, Velocity, and Acceleration (*Courtesy, Metrix Instrument Co.*)

Peak acceleration in sinusoidal vibration is defined by the following formula:

$$a = (-\omega^2 X) \sin \omega t$$

where   a = instantaneous acceleration

$-\omega X^2$ = peak acceleration

ω = vibration frequency in radians per second ($2\pi f$ radians per second in angular measure where f equals frequency)

t = time in seconds

The reader will note that a radian is equal to 57.3° angular measurement.

Vibration Classification

The three parameters (displacement, velocity, and acceleration), along with selected test frequencies, are the factors involved in vibration classification. The parameters are plotted on charts against frequency (see figure 4-2). These factors are selected so that results can be interpreted and test data will be of practical use. The graphs are illustrations of machine operating conditions.

The machine operating conditions are shown as lines labeled with letters to define severity of condition. The conditions are as follows:

AA—dangerous
A—acute fault which may deteriorate to condition AA
B—some fault
C—normal
D—faultless

It is rather obvious that the letter D represents the best condition and the levels of condition deteriorate down the alphabet to AA.

Let's consider the lower chart. This chart plots displacement in mils peak to peak against frequency. The condition plots indicate a faultless condition at 5 hertz with no displacement and dangerous conditions at frequency levels of, say, around 600 hertz and higher of 1.0-mil displacement.

The middle chart plots velocity in inches per second (ips) peak against frequency. The condition plots indicate that the condition of the machine is the same (in terms of velocity) over a large range of frequencies from 30 hertz to 1000 hertz. This range happens to be within the vibration frequency range experienced by most industrial machinery.

The upper chart plots acceleration in g's peak against frequency. It shows peak acceleration against frequency. At lower frequencies, acceleration is variable, increasing to a plateau between 1000 hertz and 4000 hertz.

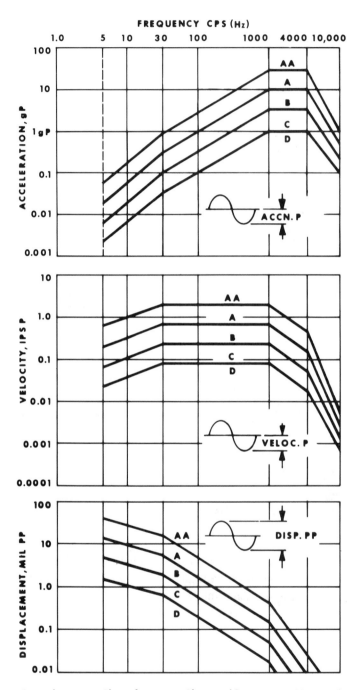

Figure 4-2 Vibration Classification Charts (*Courtesy, Metrix Instrument Co.*)

Acceleration-vibration measurements are best suited for high-frequency operation.

VIBRATION MEASUREMENT

Vibration measurement is usually accomplished by placing a transducer such as a seismic pickup or a proximity probe directly next to the machine or other equipment that is vibrating. The transducer then converts vibration to an electrical signal and feeds it to an indicating device that conditions the signal to obtain the desired measurand. For instance, acceleration can be monitored directly by feeding the signal to an ac voltmeter which is calibrated in gravity units, or the signal from a velocity pickup can easily be electrically differentiated. Displacement can likewise be measured directly with a proximity probe or by electrically integrating signals from a velocity transducer.

Some of the modern vibration analyzers are capable of providing indicator or digital readouts of displacement, velocity, and acceleration on calibrated panel meters. These same readouts may be seen on an oscilloscope for form and shape viewing.

Measurement seems to be the easy part of vibration studies. However, measurement values must be analyzed to tell the meaningful details of what the results actually say. We shall discuss some of the results in the next several paragraphs.

Amplitude

In dynamic motion (vibration) parameters, the amplitude is generally considered to be the peak value of the signal. Let's take the case of a machine vibration. Continuous monitoring of the machine will generally show that peak-to-peak displacement will be a stable amplitude and measurable. If the amplitude increases or decreases, chances are that there is justification to investigate the vibration condition further. The amplitude of vibration on most machinery is expressed in peak-to-peak mils or micrometers displacement. Displacement is usually accomplished by placing proximity probes near bearings. Vibration tolerances are established that provide for the maximum excursion of a shaft with respects to the inner race of the bearing.

Frequency

Frequency is the amount of signal cycles per time interval. For instance, house electricity is transmitted in terms of 60 cycles per second (cps) or, as it is referred to now, 60 hertz. Frequency of machine vibration is expressed in multiples of rotative speed of the machine and measured as the frequency of vibration in cycles per minute (cpm). The tendency of a machine is to vibrate at direct multiples or submultiples of the rotating speed of the machine. This makes it rather handy for persons working on the machine to express the frequency of vibration and to segregate the problems that may have occurred. Often there is a relationship between the vibration frequency, the machine's rpm, and the malfunction. For instance, the machine could have a vibration frequency that is twice that of the machine rpm. This is not usually the case, for a specific vibration frequency may include several possible problems. However, the frequency of vibration and the machine rpm are related in this manner (50 percent rpm, 33 percent rpm, etc.). This is called *synchronous vibration.* Nonsynchronous vibration occurs when the vibration is not locked in at a frequency that is a multiple of the running speed of the machine.

Phase Angle

Electrically speaking, the phase angle is the angular difference between a reference frequency and an imposed signal frequency. It is also referred to as the angular relationship between two frequencies. Phase-angle measurement is a valued test to monitor such things as the high spot of a rotor at an instantaneous point in time. By knowing the high spot, the person analyzing the condition may determine the rotor balance. Extremely large machines are designed with built-in monitors to measure frequency of vibration. Portable phase-angle instrumentation is used for monitoring smaller equipment.

VIBRATION DEVICE SELECTION AND APPLICATION

The choices of transducer type and overall response are independent decisions of the user. The decisions are usually based on frequency and the rotor/stator mass ratio. Below are some response notes directed toward correct application:

1. Use a displacement transducer, noncontact when the vibration frequency is below 1000 cycles per minute (cpm). Displacement response should be specified when a noncontact displacement sensor is employed, regardless of frequency, since velocity or acceleration readout is not practical with this type of transducer.
2. Use a displacement sensor when the rotor/stator mass ratio is small, such as in centrifugal compressors, or when large vibrations are transmitted to the base from other machinery, in which instance the use of a seismic transducer may give erroneous readings.

The acceleration sensor is used to detect high-frequency vibration, typical of gear boxes and high-speed turbo machinery. This type of sensor exceeds the frequency response of a velocity transducer. The complete acceleration sensor consists of an accelerometer, a coaxial sensor cable, and a miniature charge amplifier that provides a high-level, low-impedance output suitable for driving a long transmission cable to a vibration monitor or other receiver.

The accelerometer consists of a self-generating piezoelectric element enclosed in a stainless-steel housing with a coaxial connector. The sensing element is electrically isolated from the mounting base to prevent spurious electrical noise caused by ground loops. Mounted internally is a voltage surge arrester, which prevents the accumulation of dangerously high static electricity. The unit can be mounted directly to the machine with a stud.

Velocity Sensor (See figure 4–5)

A typical velocity sensor is the Metrix Instrument Company Model 5166 velocity transducer.

Seismic velocity transducers mount directly on a machine and measure seismic vibration—i.e., vibration relative to a fixed point in space. They convert mechanical vibration to an electrical signal that is proportional to the instantaneous vibration velocity. The transducer consists of a spring-suspended mass within a case that is surface-mounted on the machine. The mass, being

Figure 4–5 A Typical Velocity Sensory (*Courtesy, Metrix Instrument Co.*)

self-generating, requires no power source. With a relatively high output and low source impedance, the transducer may drive long transmission cables without intermediate amplification.

THE ACCELEROMETER

An accelerometer is a transducer that responds to acceleration in one or more axes. Other names for an accelerometer are *vibrometer, seismometer, velocimeter,* and *inclinometer.* These devices all have some similar basic principle involved. They are all vibration pickups consisting of a mass and a spring constant. They are most generally damped and fit into a housing of sorts. They all differ, however, in their natural frequency and the method by which motion between the mass and the housing produces an electrical signal.

Acceleration is the rate of a velocity change. Velocity is a vector quantity that involves both magnitude and direction. Velocity may change in amount and direction or in both. Acceleration, then, is also a vector quantity.

Acceleration is usually measured in feet per second^2 or meters per second^2. If a mass is moving at a constant speed, the acceleration is zero. If a mass is moving in a constant direction, the acceleration results in a continuous change in speed.

Acceleration Further Defined

There are two fundamental forms of acceleration. These are linear and angular.

Linear Acceleration   Linear acceleration occurs when all points of a rigid mass are moving in parallel straight paths. The mass is moving in a linear path. An aircraft, for instance, when it is moving in straight path, speeds up or slows down. In each case it experiences a linear acceleration. When making a bank, the aircraft changes direction and again experiences a linear acceleration.

The basic formula for linear acceleration is as follows:

$$a = \frac{\Delta V}{t}$$

where   a = linear acceleration
ΔV = change in velocity
Δt = change in time

The formula is, of course, limited to an instantaneous acceleration. If the acceleration is constant, average and instantaneous accelerations are equal.

Let's consider a mass rotating on a string tied to a fixed center (see figure 4–6). The mass undergoes a linear acceleration, considering that it constantly changes its direction of motion toward the center of the circle. This is called *centripetal* (radial) *acceleration*. In the figure, a_r is the radial component in circular motion. The vector a_r has a magnitude of V^2/R where V is velocity and R is the radius of the circular path. Radial acceleration is required to direct the mass in a circular path.

The second type of linear acceleration is *tangential acceleration*. It is represented in figure 4–6 as the vector a_t. Its magnitude is αR. The component of tangential acceleration is tangent to the path of the mass about the axis of rotation.

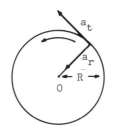

Figure 4-6   Radial and Tangential Acceleration

Angular Acceleration   Angular acceleration is a vector quantity that represents a rate of angular velocity change of a mass in a rotational motion. Consider a mass which in time t_a has an angular velocity ω_a and a later time t_b has an angular velocity ω_b. The acceleration average is thus:

$$\bar{\alpha} = \frac{\omega_b - \omega_a}{t_b - t_a} = \frac{\Delta\omega}{\Delta t}$$

These average angular accelerations are given in radians per second^2. The instantaneous angular acceleration is given by the formula:

$$\alpha = \frac{d\omega}{dt}$$

If a mass is rotating around a point with an angular acceleration of magnitude α and at an angular velocity of ω_a at a given time, then at a later time b, the angular velocity, is calculated as follows:

$$\omega = \omega_a + \alpha_b$$

A merry-go-round rotates 360° in one revolution. It has rotated through 2π radians in angular motion. The angular velocity is measured in radians per

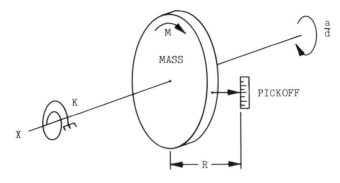

Figure 4-7   Angular Acceleration

second or degrees per second. Angular velocity is a vector quantity that has direction and magnitude. Angular acceleration is the rate of change of angular velocity.

Let's consider angular acceleration as illustrated in figure 4-7. Here X is the axis of symmetry, R is a radius between the axis of symmetry and the pickoff, K is the spring constant (as in all accelerometers), M is the seismic mass (moment of inertia), and a/d is the angular amplitude and direction. The ratio that follows is:

$$\frac{a}{d} = \frac{K}{M}$$

ACCELEROMETER BASICS

The accelerometer, as was previously described, is a device that monitors or measures acceleration.

The Seismic System (See figure 4-8)

The sensing element involved in most linear accelerometers and other vibration devices may be represented by a spring-mass combination which deviates from a fixed damped position within a frame. This follows Newton's second law of motion, $F = ma$, where F equals force, m equals mass, and a equals acceleration.

The seismic system (short for seismographic) contains a mass (seismic element) suspended from a frame with a spring. The mass is centered (guided) by a pair of spring guides. Damping of the mass is often provided by mechanical or

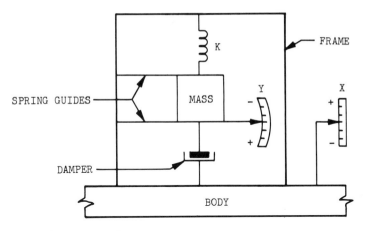

Figure 4-8 The Seismic System

electrical means. An electrical pickoff monitors the position of the mass in relation to the frame.

In the illustration, X represents movement in space of the body being monitored, Y represents the motion or movement of the suspended mass with respect to the frame, and K is the spring constant of the mass suspension.

The seismic system responds to an acceleration by producing a force proportional to the applied acceleration. An equal reaction force is developed by the spring. The deflection is a linear function of acceleration. The acceleration is held within the realm imposed by the natural frequency of the seismic system and its damping ratio.

Open-Loop Accelerometers

Often the seismic system is coupled to another tranducer outside its frame, such as a linear variable differential transformer (LVDT). The seismic system and the external LVDT combine to create a transducer whose output is proportional to the applied acceleration. This combination is called an *open-loop accelerometer*.

Natural Frequency

Natural frequency is that frequency at which a seismic system vibrates without damping. It is a measure of the speed at which the seismic system moves in response to some acceleration.

Natural frequency can be calculated by the formula:

$$2\pi f_n = \sqrt{k/m}$$

or

$$f_n = \frac{\sqrt{k/m}}{2\pi}$$

where f_n = natural frequency

k = spring constant

m = mass

These formulas are for a seismic system that is undamped.

Damping

Usually, an open-loop seismic system is damped using mechanical, fluid, or electrical means. If it is not damped, it may continue to vibrate long after the cause for the acceleration has been removed. A pictorial representation of a damper is illustrated in figure 4–7.

Damping is a ratio. Damping ratios vary so that the coefficient of damping varies between 0 and 0.7 of the critical damping required for accelerometer operation.

Accelerometer Types

The first of the accelerometer types is the variable-resistance type (see figure 4–8). The variable-resistance accelerometer is an electromechanical sensor. As the dimensions of the resistor are varied, current through the resistance varies and in turn the voltage drop across it varies. The *slide-wire potentiometer* is typical of the variable-resistance accelerometers.

A second variable-resistance accelerometer is the *strain-gage accelerometer* (see figure 4–9). There are two basic styles of strain-gage accelerometer—the *bonded strain gage* has its gage connected directly to the spring element; the *unbonded strain gage* has pretensioned strain-gage wire supporting a seismic mass. Strain-gage accelerometer types monitor accelerations up to 1000g.

In a third variable-resistance-type accelerometer, the *piezoelectric element* is used as the spring (see figure 4–10 for a functional diagram). The piezoelectric element is a crystal semiconductor whose resistance is modified by a change in applied force. It is sensitive and can monitor accelerations up to

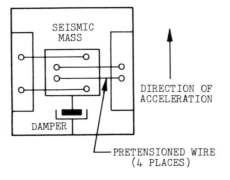

Figure 4-9   The Strain-Gage Accelerometer

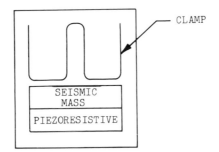

Figure 4-10   The Piezoresistive Accelerometer

10,000g. These accelerometer types are usually placed in Wheatstone bridges.

Another accelerometer is the *variable-inductance* type. This accelerometer utilizes a differential transformer with three coils. The coil in the center is excited by ac voltage. The two coils on the outside are connected in series opposition to the motion of a seismic mass passing through them. This device is used for accelerations of 50g or less (see figure 4-11).

A *piezotransistor accelerometer* contains a seismic mass tied to a stylus (see figure 4-12). The stylus transmits a force. The force causes a stress on one surface of a transistor. This surface of the transistor is part of a pn junction diode. The force causes a current change in the diode which, in turn, causes the transistor to operate. The change in current is proportional to the amount of acceleration. The piezotransistor accelerometer monitors accelerations of 400g and less.

The *servo accelerometer* contains a servo mechanism that holds a seismic mass in position as its case is accelerated. The amount of force required to keep the mass restricted is proportional to the acceleration.

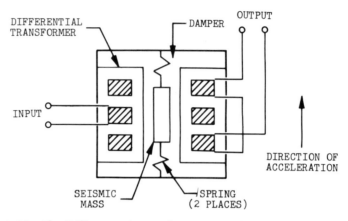

Figure 4-11   The Differential-Transformer Accelerometer

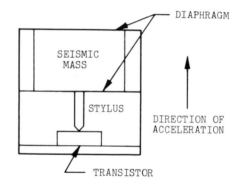

Figure 4-12   The Piezotransistor Accelerometer

One angular accelerometer has damping fluid as its seismic mass. During angular acceleration, the fluid rotates through the housing and acts on two vanes. The vanes are symmetrical. The pressure on the vanes is proportional to the acceleration.

A second angular accelerometer is a seismic mass that is mounted in a damper of fluid. The mass is mounted so that it rotates about an axis through its center of gravity. The angular deflection of the mass restricted by a spring is proportional to the acceleration.

Calibration of an Accelerometer

The performance of an accelerometer must be measured statically and dynamically. The results derived from these tests determine the response of the accelerometer. Static response checks are made for range and linearity. Dynamic response checks are made for damping and natural frequency.

Static Checks There are two basic methods for static calibration. One is the *2g turnover method*. With this method the accelerometer is placed on a level platform that may be inclined from one g to zero to one g, respectively. The difference in indications between the two points represents the basic *static* calibration.

A second method for static calibration is the *centrifuge method*. This method is used for accelerations greater than one g. The seismic mass is installed on a table of a rotary accelerator. The axis of rotation should be vertical. The response to static acceleration is thus:

$$a = 4 \pi^2 N^2 R$$

where a = acceleration in g units
N = velocity in rpm
R = radius of rotation of the center of gravity of the mass

The accuracy of the above setup is determined by the methods used to find the center of gravity of the mass and the radius of rotation.

Dynamic Check The simplest model for calibrating an accelerometer dynamically is to install it, and a precalibrated accelerometer that has the same natural frequency, on a shaker table. Then both accelerometers are dynamically shaken. The differences between the test instrument and the calibrated instrument are then compared.

Angular Acceleration Check In this check the angular accelerometer is mounted on a torsional pendulum. The pendulum and the accelerometer are started and the swings counted. The torque is the angular acceleration.

STATE-OF-THE-ART ACCELEROMETERS

The piezoelectric accelerometer is among the most popular of the accelerometer types on the market. It is also the state-of-the-art device used in vibration applications. The author has chosen this device as typical. State-of-the art devices used in flight-control applications are the servo accelerometers. There are, of course, many more different types and models of accelerometers; however, we cannot cover all of them because of space limitations.

A Typical Piezoelectric Accelerometer

Figure 4-13 shows the internal configuration of a BBN Instrument Company Series 500 piezoelectric accelerometer. When the accelerometer is mounted on a vibrating surface, the piezoelectric crystal is alternately squeezed and stretched by the inertial mass, thus generating a small voltage. The voltage generated is proportional to the acceleration over a broad frequency range.

The advantage of the piezoelectric accelerometer is that it has no moving

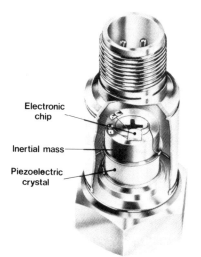

Figure 4-13 Cross-Section of a Piezoelectric Accelerometer (*Courtesy, BBN Instruments Corp.*)

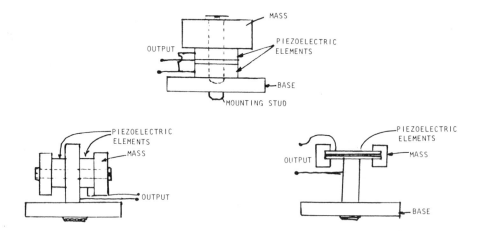

Figure 4-14 Piezoelectric Accelerometer Design Types (*Courtesy, BBN Instruments Corp.*)

parts and therefore seldom needs to be repaired or replaced. The fact that it weighs under 2 grams is also significant.

Design Types (See figure 4–14) There are three basic design types used to stress the piezoelectric element. These are compression, shear, and bending. In figure 4–14, these are illustrated top to bottom respectively.

With the *compression technique*, the mass compresses the elements that are mounted to a base. The compression technique is low cost with a high frequency range. With the *shear technique,* the elements are mounted in between the mass and in parallel with the base. The shear technique is insensitive to base bending and has good thermal characteristics.

The *bending technique* has the element isolated from the base between the mass. The bending technique has good low-frequency response and a large output.

A Typical Servo Accelerometer

Figure 4–15 is a cutaway drawing of the Sundstrand Model 303B servo accelerometer. The seismic mass is nonpendulously mounted by three flexures set at 120° apart to minimize cross axis error as well as hysteresis error. This virtually eliminates sensitivity to angular acceleration. A capacitive displacement sensor reacts to an acceleration and provides deviation signals to a servo amplifier. A feedback from the amplifier provides gain control and damping. An oscillator provides rf excitation. Gain and range resistances on the ouput legs make the unit flexible to change when desired. Current through an electrically isolated test coil applies a force to the sensing element and makes possi-

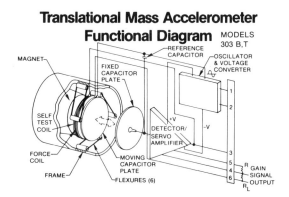

Figure 4–15 Cross-Section of a Typical Servo Accelerometer *(Courtesy, Sundstrand Data Controls Inc.)*

ble a valid check of accelerometer functioning prior to use. The self-test coil also allows nulling of the effect of earth's gravity field for low-level acceleration measurements perpendicular to the earth's surface.

REFERENCES

Reference data and illustrations were supplied by state-of-the-art manufacturers. Permission to reprint was given by the following companies:

Acceleration/displacement/velocity—Metrix Instruments Co., Houston, Texas

Vibration accelerometers—BBN Instruments, Inc., Cambridge, Massachusetts

Servo accelerometers—Sundstrand Data Control, Inc., Redmond, Washington

All copyrights © are reserved.

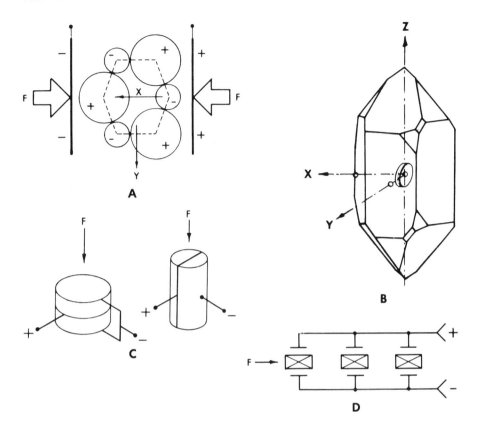

Figure 5-2 Crystal Dynamics *(Courtesy, PCB Piezotronics, Inc.)*

matically illustrates the displacement of electrical charge due to the deflection of the crystal lattice. The large circles represent silicon atoms and the smaller ones represent oxygen of a quartz crystal.

Figure 5-2A illustrates the direct effect where the force is applied along the X axis and the signal removed on surfaces perpendicular to it. It is apparent from the figure that the same deflection could be accomplished by a tension force applied along the Y axis, producing a transverse effect.

Crystalline quartz, either in its natural or reprocessed form, is one of the most sensitive and stable of piezoelectric materials. Figure 5-2B shows an X-cut disc properly oriented within a quartz crystal.

As shown in figure 5-2C, two variations of X-cut quartz elements are frequently employed in transducers. Elements may be connected electrically in parallel and mechanically either in series or parallel. The elements illustrated in figure 5-2D are mechanically in series and electrically connected in parallel to produce more charge output.

The Crystal Transducer Function (See figure 5–3)

The crystal elements in piezoelectric transducers perform a dual function. They act as a precision spring to oppose the applied pressure or force and supply an electrical signal proportional to their deflection. The physical configuration of pressure, force, and acceleration transducers employing transverse elements are shown in figure 5–3. Note that they differ little in internal configuration. In accelerometers, to measure motion, the invariant seismic mass, m, is forced by the crystals to follow the motion of the base (or structure to which the base is attached). This is an implementation of Newton's law, which states that force and acceleration are proportional.

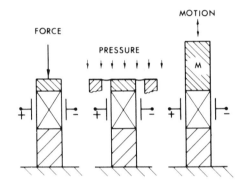

Figure 5–3   The Crystal Transducer Function *(Courtesy, PCB Piezotronics, Inc.)*

Basic Design Configurations (See figure 5–4)

The figure illustrates design setups of crystals in different configurations. As stated previously, to generate a useful output signal (− and +) crystals rely on the piezoelectric effect. When the crystal elements are strained by an external force, displaced electrical charges accumulate on opposing surfaces. A voltage signal is formed according to the law of electrostatics. In figure 5–4A, the configuration is used as a basic force/impact transducer. Figure 5–4B illustrates an acceleration-compensated pressure transducer or microphone. Further configurations are the quartz accelerometer for shock and vibration in figure 5–4C and the universal transducer (mechanical-impedance) shown in figure 5–4D. Because of similarity, transducers designed to measure one specific input are also somewhat sensitive to other inputs. Modern sophisticated transducers employ various compensating techniques to reduce this interaction. Eliminating or reducing temperature and strain sensitivity also calls for innovative design.

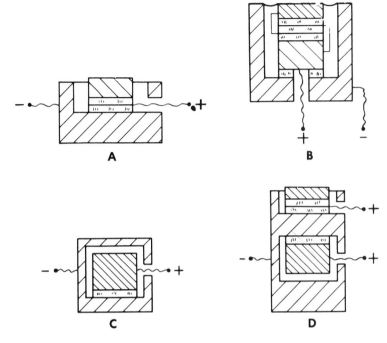

Figure 5–4 Basic Crystal Design Configurations
(Courtesy, PCB Piezotronics, Inc.)

QUARTZ

Quartz is made up of two of the most common elements found on earth, silicon and oxygen. Quartz is found in silica sand, flint, agate, and amethyst. Pure electronic-grade quartz is a rarity. Crystalline quartz is a critical defense metal. It is used in radio, radar, and navigational equipment. The United States government, under a sponsored program, has synthesized it. In an autoclave under high pressure and temperature it takes around a month to duplicate crystals that it takes nature a million years to grow. Quartz of this type is a near-perfect transduction element.

A QUARTZ FORCE TRANSDUCER

Typical of the quartz force transducers is the PCB Piezotronics Series 208A (see figure 5–5A). This force transducer measures dynamic and short-term static forces from 1.0 to 500 pounds at any tare level within this range. The structure of the transducer contains two thin quartz discs or plates operating in a thickness-compression mode and capped by hardened steel cylindrical

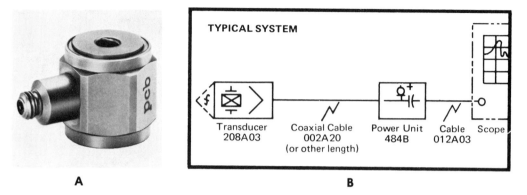

A   B

Figure 5–5   A Quartz Force Transducer *(Courtesy, PCB Piezotronics, Inc.)*

members. Stiffness approaches that of a solid steel cylinder and has little effect on the rigidity of most test structures or objects.

In figure 5–5B, the PCB Model 208A03 is placed in a typical system. When connected to a power unit, the self-amplifying transducer generates a high-level, low-impedance analog output signal proportional to the measurand and compatible with most readout instruments. The power-unit circuit powers the transducer over the signal lead (coaxial or two-wire), eliminates bias on the output, and monitors normal or faulty operation.

The unit is used for measuring compression, tension, impact, reaction, and actuation forces in testing, vibrating, tensioning, balancing, striking, welding, rolling, cutting, coining, forming, pressing, machining, and punching operations.

A QUARTZ ACCELEROMETER

A quartz accelerometer is shown in figure 5–6A. It is structured with high-sensitivity (100mV/g) permanently polarized quartz elements.

Quartz accelerometers function to transfer the acceleration aspect of vibratory motion into an electrical signal for displaying, controlling, or processing. Integrators are often employed in readout or analyzing instruments to compute velocity and displacement. Quartz measures over a wide frequency range down to one hertz.

The structure of this solid-state instrument includes a rigid compression-mode quartz and steel seismic assembly plus a built-in microelectronic amplifier, shock-protected by an electronic switch. There are no moving parts. It mounts to the test object with a threaded stud or magnetic base.

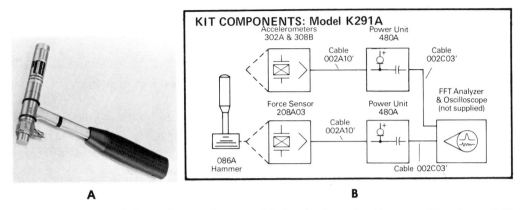

Figure 5-8 A Quartz Structural Behavior Impulse Hammer *(Courtesy, PCB Piezotronics, Inc.)*

(hammer) connects by cable to an FFT analyzer through a power unit. Accelerometers also connect to the FFT analyzer by cable through a power supply.

PIEZOELECTRIC CERAMICS

An extremely complex manufacturing process with heat and pressure provides the transducer field with synthetic crystals. These crystals, primarily made from barium titanate and lead titanate-lead zirconate are manufactured as large as 30 tons and small enough to be fed through a tiny vein to the heart.

The piezoelectric ceramic is an electromechanical device. Piezoelectric ceramics are isotropic before poling. *Isotropic* means that the material has the same properties regardless of the direction of measurement. *Poling* is a procedure that involves momentary application of a strong direct current. After poling, the ceramic becomes *anisotropic* (varies according to the direction of measurement). That is, their electromechanical properties differ for electrical or mechanical excitation along different directions. If the crystal's unit cells have no center of symmetry, they will become piezoelectric after poling. If they do have a center of symmetry, they will be inert when they are excited electrically or mechanically.

A ceramic is composed of many crystals in random orientation, each unit cell containing a dipole. By application of electrodes and a strong dc field, the dipoles are aligned parallel to the field, thus making the ceramic piezoelectric. Not every domain aligns its dipoles, but enough of them do to achieve piezoelectricity. Once polarized, the ceramic takes on its own personality, exhibiting specific electrical and physical properties to be discussed later in this chapter. Polarization is the last step, other than testing, in the manufacture of ceramics.

Modes of Vibration of the Piezoelectric Resonator (See figure 5-9)

There are three primary concerns in considering the modes of vibration of the piezoelectric resonator. These are the *capacitance of the material,* the *relative permitivity* (dielectric constant), and the *frequency constant*. These three really determine the crystal's operation. Each shape has its specific mode. Each is electroded in a particular manner, such as on flat surfaces, on ends, or on curved surfaces. Each is poled for the best possible alignment of the dipoles, which is what makes the ceramic piezoelectric. The piezoelectric industry uses the orthogonal axis originally assigned by the crystallographers to determine the piezoelectric modes of operation. That is: 1 corresponds to the X axis, 2 corresponds to the Y axis, and 3 corresponds to the Z axis. The direction of polarization is defined as the 3 axis. Table 5-1 lists the symbols used in determining mode capacitance, relative permitivity, and the frequency constants. Figure 5-9 shows the modes of vibration of the piezoelectric resonators. Each of the modes of vibration shown in figure 5-9 has material constants provided by the manufacturer which designate the specific properties of the ceramic, such as the coefficient of coupling, stress, and strain and the direction of poling. Table 5-2 lists these constants relating to the vibration mode as illustrated in figure 5-9. Table 5-3 describes the constants and their relationship to the modes of vibration.

Ceramic Materials and Their Applications

One of the most important attributes of the ceramic piezoelectric material is that each composition can be manufactured for some desirable characteristic or mode of operation. For instance, one material form is a lead titanate zirconate composition developed for high-power acoustic projectors. This material has a high coercive field, high coupling coefficient, and low dielectric losses under high driving fields. It is ideally suited for high-power, low-frequency broadband projectors and other high-power electroacoustic devices.

A second composition is a lead titanate zirconate material modified to provide an extremely high dielectric constant and a high coupling coefficient. The characteristics of this material make it ideal for hydrophones or low-power projectors requiring a high dielectric constant or ''d'' constant.

TABLE 5-1 Symbols Used for Modes of Vibration of Piezoelectric Resonators

SYMBOL		DEFINITION	UNIT OF MEASURE
C_e	=	Electrical Capacitance	Farads
C_o	=	Clamped (low frequency) capacitance	Farads
Dia	=	Diameter	Inches, Meters
d_m	=	Mean Diameter	Inches, Meters
f_a	=	Antiresonant frequency	Hertz
f_c	=	Frequency Constant	Kilohertz Meters
f_r	=	Resonant Frequency	Hertz
ID	=	Inside Diameter	Inches, Meters
L	=	Length	Inches, Meters
OD	=	Outside Diameter	Inches, Meters
r	=	Resitance	Ohms
S	=	Compliance	M /Newton
t	=	(th) Thickness	Inches, Meters
$\tan\delta$	=	Dissipation Factor	%
V	=	Voltage	Volts
W	=	Width	Inches, Meters
Y_a	=	Admittance at antiresonance	mhos
Y_r	=	Admittance at resonance	mhos
$\mathcal{E}_o$	=	Permitivity of Free Space $(= 8.85 \times 10^{-12})$	Farad/Meter
$\mathcal{E}_r, K$	=	Relative Permitivity, Dielectric Constant	
ρ	=	Density	Kg/Meter^3

Courtesy EDO Western Corp.

TABLE 5-2 Material Constants for the Various Vibratory Modes

	COUPLING COEF.	g CONSTANT	d CONSTANT	STIFFNESS CONSTANT
1. Thickness Poled	k_p	g_{31}	d_{31}	y_{11}
2. Thickness Poled	k_{33}	g_{33}	d_{33}	y_{33}
3. Thickness Poled	k_{31}	g_{31}	d_{31}	y_{11}
4. Tangentially Poled	k_{33}	g_{33}	d_{33}	y_{33}
5. Length Poled	k_{33}	g_{33}	d_{33}	y_{33}
6. Thickness Poled	k_{33}	g_{33}	d_{33}	y_{33}
7. Thickness Poled	k_p	g_{31}	d_{31}	y_{11}
8. Thickness Poled	k_{15}	g_{15}	d_{15}	y_{44}

Courtesy EDO Western Corp.

ABINGDON COLLEGE OF FURTHER EDUCATION

Library

1. Thin Disc
Radial Mode

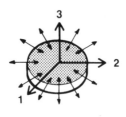

Electroded on flat surfaces.
Poled through thickness.
Frequency constant $N_p = dia \times fr$

$$Capacitance = \frac{dia^2 \times \mathcal{E}r}{5.664 \times th}$$

$$\mathcal{E}r = \frac{(5.664)(Co)(th)}{dia^2}$$

2. Thin Disc or Plate
Thickness Mode

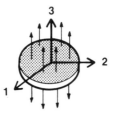

Electroded on flat surfaces.
Poled through thickness.
Frequency constant $N = th \times fr$

$$Capacitance = \frac{dia^2 \times \mathcal{E}r}{5.664 \times th}$$

$$\mathcal{E}r = \frac{(5.664)(Co)(th)}{dia}$$

3. Long thin Bar
Length Mode

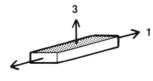

Electroded on shaded surfaces.
Poled through thickness.
Frequency constant $N = L \times fr$

$$Capacitance = \frac{\mathcal{E}r \times area}{4.448 \times th}$$

$$\mathcal{E}r = \frac{(4.448)(Co)(th)}{area}$$

4. Thin wall Tube
Radial (hoop) Mode

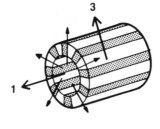

Electroded on shaded stripes.
Poled between stripes.

Frequency constant $N_3 = mean\ dia \times fr$

$$Capacitance = \frac{\mathcal{E}r \times area}{4.448 \times th \times .8}$$

$$\mathcal{E}r = \frac{4.448\ (Co)\ (ave.\ space\ between\ stripe)}{(wall)(L)(N)}$$

$N = number\ of\ stripes$

Figure 5-9   Modes of Vibration of the Piezoelectronic Resonator *(Courtesy, EDO Western Corp.)*

**5. Thin wall Tube
Length Mode**

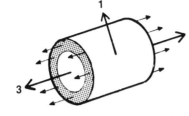

Electroded on ends
Poled through length.
Frequency constant $N = L \times fr$

$$\text{Capacitance} = \frac{(OD - ID) \times \mathcal{E}r}{5.664 \times th}$$

**6. Thin wall Tube
Thickness Mode**

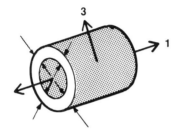

Electroded on curved surfaces.
Poled through wall thickness.
Frequency constant $N = th \times fr$

$$\text{Capacitance} = \frac{\mathcal{E}r \times L}{1.628 \times (\log 10 \frac{OD}{ID})}$$

$$\mathcal{E}r = \frac{(1.628)(Co)(\log 10 \frac{OD}{ID})}{L}$$

**7. Thin wall Sphere
Radial Mode**

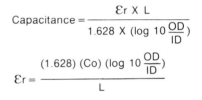

Electroded on curved surfaces.
Poled through wall thickness.
Frequency constant $N_s = $ mean dia $\times fr$

8. Shear Plate

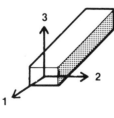

Electroded on shaded surfaces.
Poled through thickness. (3 axis)
Frequency constant $N_s = th \times fr$

$$\text{Capacitance} = \frac{\mathcal{E}r \times area}{4.448 \times th}$$

$$\mathcal{E}r = \frac{(4.448)(Co)(th)}{area}$$

Figure 5-9 *continued*

TABLE 5-3 Property Symbols and Definitions (Reference Table 5-2)

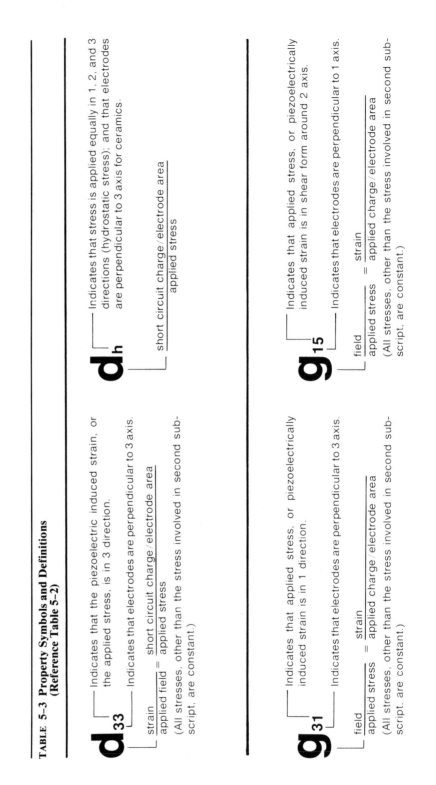

d_{33}

- Indicates that the piezoelectric induced strain, or the applied stress, is in 3 direction.
- Indicates that electrodes are perpendicular to 3 axis.

$$\frac{\text{strain}}{\text{applied field}} = \frac{\text{short circuit charge/electrode area}}{\text{applied stress}}$$

(All stresses, other than the stress involved in second subscript, are constant.)

g_{31}

- Indicates that applied stress, or piezoelectrically induced strain is in 1 direction.
- Indicates that electrodes are perpendicular to 3 axis.

$$\frac{\text{field}}{\text{applied stress}} = \frac{\text{strain}}{\text{applied charge/electrode area}}$$

(All stresses, other than the stress involved in second subscript, are constant.)

d_h

- Indicates that stress is applied equally in 1, 2, and 3 directions (hydrostatic stress); and that electrodes are perpendicular to 3 axis for ceramics.

$$\frac{\text{short circuit charge/electrode area}}{\text{applied stress}}$$

g_{15}

- Indicates that applied stress, or piezoelectrically induced strain is in shear form around 2 axis.
- Indicates that electrodes are perpendicular to 1 axis.

$$\frac{\text{field}}{\text{applied stress}} = \frac{\text{strain}}{\text{applied charge/electrode area}}$$

(All stresses, other than the stress involved in second subscript, are constant.)

128

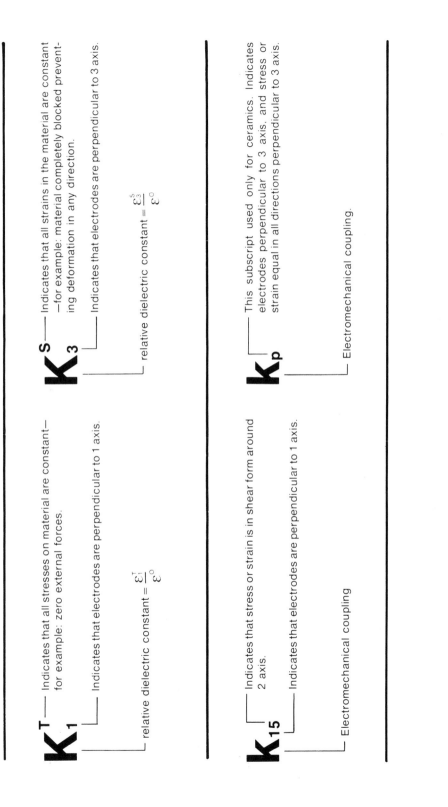

K_1^T —— Indicates that all stresses on material are constant—for example: zero external forces.

—— Indicates that electrodes are perpendicular to 1 axis.

—— relative dielectric constant = $\dfrac{\varepsilon_1^T}{\varepsilon^o}$

K_3^S —— Indicates that all strains in the material are constant—for example: material completely blocked preventing deformation in any direction.

—— Indicates that electrodes are perpendicular to 3 axis.

—— relative dielectric constant = $\dfrac{\varepsilon_3^S}{\varepsilon^o}$

K_{15} —— Indicates that stress or strain is in shear form around 2 axis.

—— Indicates that electrodes are perpendicular to 1 axis.

—— Electromechanical coupling

K_p —— This subscript used only for ceramics. Indicates electrodes perpendicular to 3 axis, and stress or strain equal in all directions perpendicular to 3 axis.

—— Electromechanical coupling.

A third material is a modified barium titanate composition that has been developed to provide an extremely low-loss material with excellent stability of parameters from $-40°$ C to $100°$ C. This material has ideal characteristics for high-power ultrasonic cleaning systems and low-loss highly stable sonar projectors.

Of course there are many other compositions that can be manufactured. Crystals in their natural state do not have this flexibility.

Other Ceramic Characteristics

As you may have determined, there are characteristics of ceramics that define either a range of operation or a parameter that must be met.

One of the most important of the specifications of the ceramic is the temperature performance curves. These curves compare such things as dielectric constant versus temperature, frequency constant versus temperature, and coupling coefficient versus temperature.

Mechanical tolerances such as inside or outside diameters, lengths, and flatness are other specifications.

The ceramic's electrode characteristics are usually provided. There are two types of electrodes common to the piezoelectric industry. These are silver and electroless nickel. Other materials such as copper or electroplated nickel are also available. An electrode's qualities are best defined in terms of adhesion, conductivity, cosmetic appearance, and surface finish.

Evaluation of the characteristics of the piezoelectric resonator is dependent on accurate measurements of the physical and dielectric properties and impedance of the element or resonator. While the measurements of density, size, clamped capacitance, and low-field dissipation factor are relatively simple, the evaluation of the element impedance and critical frequencies must be carefully performed.

An Acoustical (Ceramic) Transducer

An example of an acoustical (ceramic) transducer is the EDO Western Corporation Model 6194 (see figure 5–10).

This transducer is designed for making sound-pressure measurements over the frequency band of 150Hz to 100kHz.

The unit contains an integrated-circuit preamplifier with a high-input impedance allowing a high sensitivity and a low-frequency response. The low output impedance of the preamplifier drives long cables with negligible losses. An electrostatic shield surrounding the preamplifier assembly is also included to provide an acceptable signal-to-noise ratio.

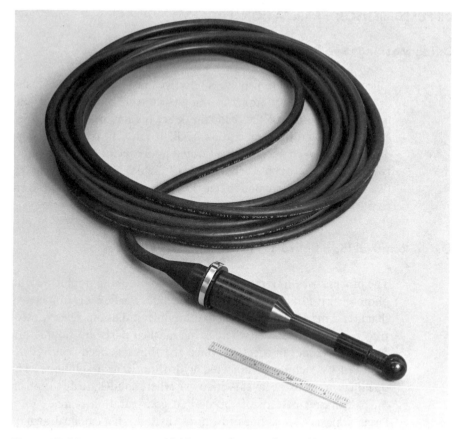

Figure 5-10   An Acoustical (Ceramic) Transducer *(Courtesy, EDO Western Corp.)*

INTEGRATED SENSORS

The demand for smaller transducers has resulted in the introduction of miniature and microminiature transducers. These sensors are made from solid-state materials and are called integrated sensors (IS). Two basic sensor series of this type are made by Kulite Semiconductor Products, Inc. These are the diffused sensors for low temperatures (up to 325°F) and the dielectrically isolated sensors for high temperatures.

A third type is the integrated-circuit piezoelectric (ICP) transducer built by PCB Piezotronics. There are of course others in this pioneering direction, but we do not have room in the book to cover them all.

DIFFUSED SENSOR FABRICATION PROCESS

Oxide Masking (See figure 5-11A)

Silicon oxide is used to provide a selective mask against the diffusion of donor or acceptor impurity atoms in this device technology. Factors affecting the masking properties of the oxide mask are the impurity source, the carrier-gas atmosphere, and the temperature of diffusion. The oxide layer is produced in dry or wet oxygen and water-saturated nonreducing gases at temperatures between 800°C and 1000°C. The range of the oxide thickness found to be effective in masking is from one to five thousand angstroms. An angstrom, you may recall, is one hundred millionth of a centimeter.

Oxide Removal (See figure 5-11B)

The process used to remove the oxide is a photolithographic technique. A thin photosensitive film is spread over the oxide. Part of this film is masked. Parts that aren't masked are exposed to ultraviolet radiation. The exposed part of the film becomes insoluble, whereas the masked parts are soluble in developing fluid. The oxide, in areas in which the film has been removed, can now be removed by etching in buffered hydrofluoric acid solutions. This provides openings in the oxide layer (windows), where solid-state diffusion can proceed.

A master pattern of the desired configuration is made on a relatively larger scale as a set of line drawings. Anything that can be drawn on a line can be reproduced. Extreme accuracy results because the patterns can be reduced photographically up to 1000 times. The fineness of the detail is in the order of resolution of the microscope. Line widths of 0.0001 inch are common in practice. This oxide-removal method is also desirable because it leaves a surface that is not strained mechanically, as would be the case with grinding, sawing, or scratching through the oxide film.

Diffusion (See figure 5-11C)

The strain-sensing network on the silicon crystal is formed through the windows in the silicon oxide mask. To form a doped layer by solid-state diffusion, a single-crystal material is heated in the presence of vapor of the desired impurity. The vapor pressure and the semiconductor temperature determine the impurity distribution. For fabrication of devices, it is important to control the depth of the diffused region and the distribution of the impurities within it. The

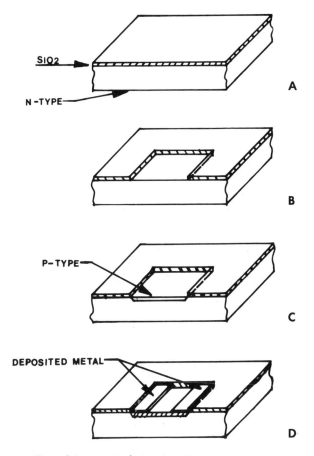

Figure 5-11 Diffused Sensor Fabrication Process

device designer can control this by varying time, temperature, and surface concentration.

Contact Metallization (See figure 5-11D)

There are two basic methods of metallizing contact lands. The first is *aluminizing,* the second is *plating of gold-nickel thin film.*

In the first method, aluminum is evaporated over the surface and preferentially removed. To ensure a good electrical contact, all traces of the silicon oxide must be removed prior to evaporation of the aluminum. Layer thickness is in the order of 10,000 angstroms. Layers of this type are suitable for thermocompression bonding. Current practice includes platinum, gold-chrome, and other materials. Evaporation, sputtering, and electron-beam deposition are current deposition techniques.

The second method involves simultaneous electroless plating of a gold-nickel film. Oxide is removed preferentially and the plating occurs only on bare silicon. Thicker films are available that make this method suitable for lead attachment by thermocompression, bonding, microwelding, or microsoldering. This method also eliminates the possibility of gold-aluminum intermetallic compound formation (purple plague). This can occur in a thermocompression bond of gold wire to an aluminum layer, where thin layers of aluminum are employed.

JUNCTION ISOLATION BY DIELECTRIC LAYERS AND EPITAXIAL GROWTH

The purpose of this method is to be able to choose the resistivity of the single crystal so as to provide the optimum thermal and electromechanical properties for each application.

Oxide Masking (See figure 5–12A)

The starting material is a properly oriented single crystal of the desired resistivity. The temperature coefficient of the gage factor and resistance are kept as low as possible. Variations of temperature are critical. This starting material is in the form of a wafer with its surfaces etched and polished to a mirror finish. The wafer is in the form of a disc, 1 inch in diameter and 0.003 to 0.005 inch thick. The wafer size will permit the simultaneous fabrication of sensors on a single wafer.

In practice, a layer of silicon dioxide 10,000 to 15,000 angstroms thick is thermally grown on the surface of the wafer. A layer such as this can be grown by heating the wafer at 1100°C to 1200°C in an atmosphere of steam for 4 to 5 hours.

Channel Etching (See figure 5–12B)

After removal of the photoresist used to remove silicon oxide from predetermined areas, the wafer is chemically etched in a composition of nitric and hydrofluoric acids that attacks the silicon and not the silicon oxide. Channels are etched in the silicon to depths of approximately one mil, controllable to $\pm$ 10 percent. Masked areas narrower than 0.005 inch can be maintained by use of proper masks. With this technique, a great number of sensors can be made on a single structure. Next, a new layer of silicon oxide is grown on the surface, resulting in the structure shown in figure 5–12B. The second oxide serves to

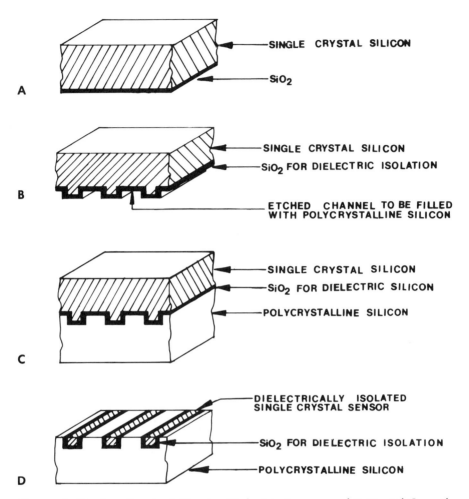

Figure 5-12 Junction Isolation by Dielectric Layers and Epitaxial Growth *(Courtesy, Kulite Semiconductor Products, Inc.)*

dielectrically isolate the sensors from the polycrystalline layer that will subsequently be grown.

Epitaxial Growth of Polycrystalline Layer (See figure 5-12C)

A polycrystalline layer of silicon is grown on top of the silicon oxide to a thickness of 0.005 inch to 0.006 inch. The channels previously etched are also filled. The new deposited area is polycrystalline because it is deposited on a silicon oxide surface. Single-crystal epitaxial layers can be grown on a single-crystal silicon substrate. The purpose of the polycrystalline layer is to provide a support and serve as a mechanical flexure for the single-crystal regions that are to be used as strain sensors in an electromechanical transducer.

Removal of Single-Crystal Material (See figure 5–12D)

The original single-crystal substrate is lapped and polished down to the level of the oxide in the bottom of the channels. Further careful polishing will remove the silicon oxide from the top surface, exposing the underlying polycrystalline silicon. This step isolates the single-crystal silicon regions which are now the piezoresistive elements. Further material removal will serve to reduce the sensor thickness without affecting the geometry of the structure. This allows the designer to tailor the sensor resistance to the desired level.

Contact areas are defined by reoxidizing and subsequently photoresisting and then preferentially metallizing the contacts. Prior to lead attachment, the individual sensors are separated from the wafer. A completed sensor (in this case an integral diaphragm) is shown in figure 5–13.

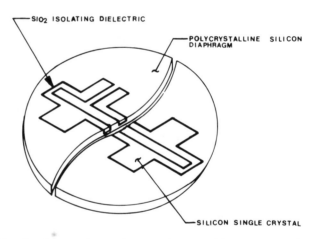

Figure 5–13 An Integral Diaphragm Sensor (*Courtesy, Kulite Semiconductor Products, Inc.*)

Integral Silicon Diaphragms (See figure 5–14)

Using the techniques of fabrication just described, integral silicon diaphragms can be made incorporating a four active arm Wheatstone bridge. The bridge provides an output that is proportional to pressure and/or deflection. The stress sensors are arranged so that under load, two elements are in tension and

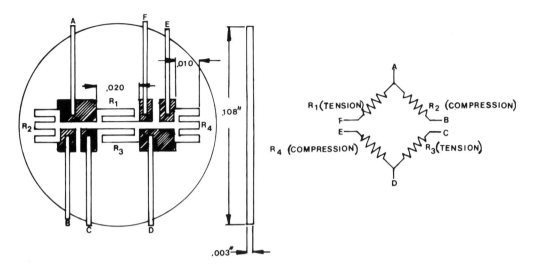

Figure 5–14 An Integral Silicon Diaphragm Sensor Containing a Four Active Wheatstone Bridge *(Courtesy, Kulite Semiconductor Products, Inc.)*

two are in compression. The gage arrangement is shown in figure 5–14. The overall area of the diaphragm is 0.108 inch. The active area is 0.090 inch. The diaphragm may be clamped on its rim using a rim thickness of 0.010 inch. The figure illustrates inner gages at 0.020 inch and outer gages at 0.010 inch. Lead contacts are made at the region of minimum diaphragm stress. Diaphragms may be made between 0.001 inch and 0.010 inch. The thickness determines the rated load and output.

The use of a dielectrically isolated integral silicon diaphragm as the force collector makes possible the construction of transducers of extremely high natural frequencies capable of high-temperature operation.

Transducer Construction (See figure 5–15)

The sensor itself does not make a transducer. Just as an integrated-sensor approach to the sensing element made possible sensor-size reduction, a transducer has to be developed to withstand the aerospace environment. Figure 5–15 describes the construction of the Q-type pressure transducer.

This type of transducer is designed to eliminate any internal unsupported leads and to avoid lead flexing. Strain-gage terminals are an integral part of the pressure diaphragm. Metallized channels extend to the clamped portion of the diaphragm where lead connection is made and secured by potting.

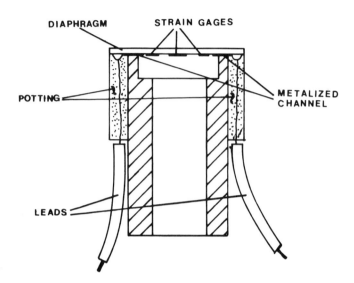

Figure 5-15 The Q Transducer Construction *(Courtesy, Kulite Semicon-*ductor Products, Inc.)

Transducer Protection (See figure 5-16)

When used in the air stream, the pressure transducer must be protected from particle impingement. Figure 5-16 describes some of the screen configurations that are currently used. In all cases, a perforated protective shield is installed in front of the diaphragm.

Besides being an effective protection against particle impingement, recent tests indicate that the screen (and especially the b configuration) acts as an excellent heat shield.

The d configuration shows a recessed screen. The cavity in front of the screen serves as the collector which, in conjunction with the chamfered edge, improves performance for variations in flow angularity. The screen does not alter the frequency response of the transducer over a range of 0 to 20kHz.

Typical Integrated Sensor

Typical of integrated sensors is the Kulite Ultra Miniature Pressure Sensor and Pressure Probe illustrated in figure 5-17. This device has wide application in wind-tunnel-engine and flight-test work such as for total pressure measurement in supersonic wind tunnels.

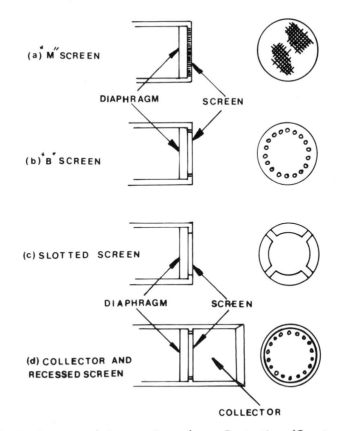

(a) M" SCREEN

DIAPHRAGM   SCREEN

(b) B' SCREEN

(c) SLOTTED SCREEN

DIAPHRAGM   SCREEN

(d) COLLECTOR AND
RECESSED SCREEN

COLLECTOR

Figure 5-16 Integrated Sensor Transducer Protection *(Courtesy, Kulite Semiconductor Products, Inc.)*

The diaphragm is usually protected by one of two methods. In the first, a perforated screen is used to shield the diaphragm. In the second, a coating is formed over the diaphragm with a protective layer such as silicone rubber.

In most cases both a protective layer and screen are utilized.

There are two screen types used on the Q transducers. One screen consists of a thick, fine metallic mesh. The second screen consists of a thick plate with holes positioned on a circle. The diameter of the circle is greater than the active diameter of the diaphragm. This arrangement eliminates any possibility of a particle penetrating through the holes and hitting the unclamped portion of the diaphragm.

Either screen is mounted in a screen holder, which is then installed on the transducer housing in front of the diaphragm.

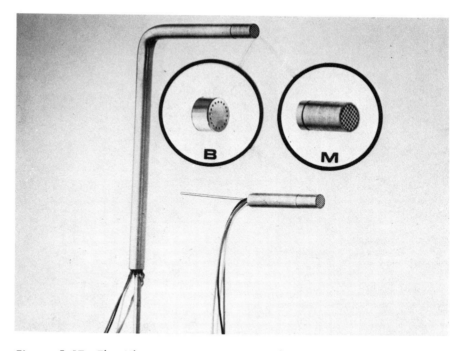

Figure 5–17 The Ultra Miniature Integrated Sensor Used as a Pressure Transducer *(Courtesy, Kulite Semiconductor Products, Inc.)*

The Integrated-Circuit Piezoelectric (ICP) Instrumentation System

The ICP system involves the combination of piezoelectric transducer and amplifier within the same package. Figure 5–18A shows a circuit diagram of the ICP transducer. The element is crystalline quartz, while the amplifier is typically a P-channel MOSFET source follower with unity gain. The measurand acts on the quartz element. The element produces a quantity of charge (q). The charge collects in the shunt capacitor (C) forming voltage (V) equal to $\frac{q}{c}$. Almost instantaneously the voltage (V) appears at the output of the amplifier, where it is added to the dc voltage level at that point. It must be remembered that the output impedance is very low and compatible with almost any readout.

Figure 5–18B illustrates the internal configuration of an ICP amplifier. Parts can be compared with the circuit diagram in figure 5–18A. Note that the source resistor can be placed anywhere along the signal lead.

Figure 5–18C is representative of a typical hookup with the resistor (R) in two places. Power is supplied by battery (B), capacitor (C) is cable impedance, and the readout is an oscilloscope.

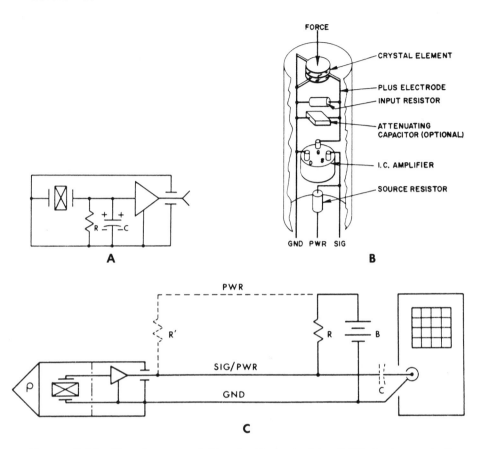

Figure 5–18 The Integrated-Circuit Piezoelectric (ICP) Instrumentation System *(Courtesy, PCB Piezotronics, Inc.)*

REFERENCES

Reference data and illustrations were supplied by state-of-the-art manufacturers. Permission to reprint was given by the following companies:

Bell and Howell, CEC Division, Pasadena, California.
EDO Western Corp., Salt Lake City, Utah.
Kulite Semiconductor Products, Inc., Richfield, New Jersey.
PCB Piezotronics Inc., Buffalo, New York.
All copyrights © are reserved.

6

Machine Monitoring: Proximity and Power

PROXIMITY DETECTORS

Proximity detectors are electical or electronic sensors that respond to the presence of a material. This electrical or electronic response is utilized to activate a relay or switch and/or to perform automation functions. Proximity detectors act as sensing devices in applications such as limit switches, liquid-quantity/level controls, speed controls, counters, and inspection tools. There are many types of proximity detectors. The major types are inductive, magnetic, and capacitive. The inductive and magnetic sensors require that the monitored material be metal. The capacitive proximity sensor can monitor nonmetal materials.

For the sake of review, let us take a second look at these transducer types. In the *inductive transducer element,* the inductance of the element is changed by the proximity of metal materials. In the *magnetic transducer element,* the magnetic field is changed by the presence of metal materials. In the *capacitive transducer element,* the capacitance of the transducer element can be changed by the presence of nonmetallic materials. The eddy-current transducer uses the principle of impedance variation. An eddy current is induced into a conductive metal target and monitored by a sensor. The temperature transducer monitors temperature with a sensitive conductive thermopile.

Proximity Defined (See figure 6-1)

Proximity is defined by Webster as the state or quality of being near. In transducer applications, it is this nearness that allows the change in electrical or

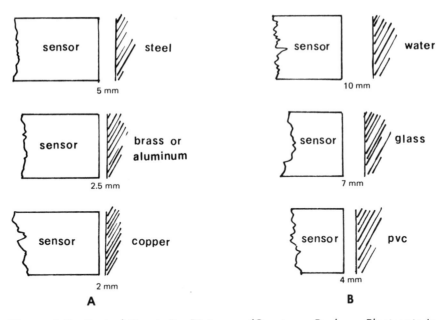

Figure 6-1 Typical Proximity Distances *(Courtesy, Rechner Electronic Industries, Inc.)*

electronic function to take place. In the figure 6-1A, some typical proximity distances for metallic materials are shown. In figure 6-1B, some typical proximity distances for nonmetallic materials are shown.

Proximity detectors have no moving parts and therefore do not wear out from constant use. Since they do not contact their target materials, they are not destroyed by rough or abrasive parts. Proximity detectors can sense small, lightweight parts without detaining them and delicate, painted, or polished surfaces without marring them. They can detect irregular-shaped objects regardless of the direction of entry into the sensing field. Control functions are possible at electronic speed.

Mounting Proximity Detectors

One of the extreme advantages of using the proximity detector is its versatility in mounting. It may be mounted external to the material being monitored, internally within the material, on flush mounts, or on nonflush mounts. In other words, the proximity detector may be mounted, depending on the device function, for material detection whether fixed or moving.

Selection of a Proximity Detector

Several variables must be considered in choosing the correct proximity detector regardless of type. These are as follows:

1. Determine the type of material that is to be monitored.
2. Determine the type of proximity detector that will best monitor that material.
3. Determine the space available to mount the detector. If the detector face will be mounted with the side of the case near or touching the machine body, choose a sensor that can be mounted flush in the metal. Note that inductive sensors can touch nonmetallic material without producing a false signal. Note that capacitive sensors will detect almost anything in the measurement path. Note also that the larger the detector, the greater the detection distance.
4. Determine how far the detector must be mounted from the target material. Select a sensor that can detect a greater distance than required.
5. Determine the amount of power required and secure a power supply that will provide that amount with a sufficient added power capability to withstand minor current surges.
6. Determine whether special mounts are required for problems such as vibration.
7. Determine matching impedances of readout devices, electronics, relays, and/or other connecting apparatus.

The Inductive Proximity Detector

As was previously stated, the inductive sensor element undergoes an inductance change owing to the proximity of material. This is usually accomplished by the inductance ratio of a pair of coils in the transducer.

Other inductive sensors project a very low-level inductive field in front of the sensor. As a conductive target enters the field, eddy currents are generated in the target that reduce the impedance of the sensor.

Inductive Sensor Operating Points   The operating point of the inductive proximity sensor is necessarily different for all materials and, of course, all sensors. Each manufacturer of sensors provides a curve that describes how much of the sensor must be covered to produce an output signal. An example of this curve is illustrated in figure 6–2. Let's consider, for example, a steel target whose sensing distance is 8 millimeters. In order for the sensor to begin operation, 16

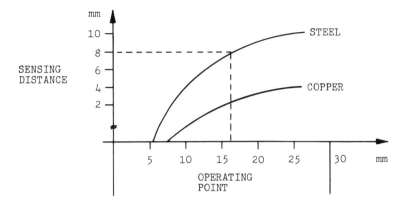

Figure 6-2 Operating Points Versus Sensing Distance of an Inductive Prox-imity Detector (Courtesy, Rechner Electronics Industries, Inc.)

millimeters of the sensor would have to be covered. Each sensor has a material scale factor. These scale factors are determined by the sensor manufacturers. Scales factor for this set of curves as follows:

Material	Scale Factor
Steel	1.0
Chrome-Nickel	0.85
Stainless	0.7
Brass	0.45
Aluminum	0.4
Copper	0.3

To determine sensing distance for other metals using the same sensor, simply multiply the sensing distance for steel times the scale factor.

Example: 8 mm × 0.3 = 2.4 mm

If copper covered 16 millimeters of the sensor it would be detected at 2.4 millimeters.

A Typical Inductive Proximity Sensor

Typical of these sensors is the Rechner IAS 20. The sensor is illustrated in figure 6-3. This sensor operates on dc voltage 10/30 vdc. Up to six of the sensors can be wired in series or parallel to perform logic functions.

Figure 6-3 An Inductive Proximity Detector (*Industries, Inc. Courtesy, Rechner Electronics*)

Inductive Proximity Detector Applications

The uses of this device are numerous. Two of these applications are illustrated in figure 6-4. Inductive sensors can be used for measuring metallic materials. Flush-mounted sensors of the detector have a direction field that extends in front of the sensor head. The field does not extend beyond the side of the sensor. The well-defined sensing field allows the detection of closely spaced items and produces one specific signal for each item. To accurately distinguish parts in a row, the sensor diameter must be no greater than the narrowest space between objects.

Web-break detection can be accomplished with inductive sensors as

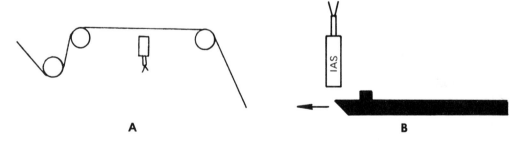

Figure 6-4 Applications for Inductive Proximity Detectors (*Courtesy, Rechner Electronics Industries, Inc.*)

shown in figure 6–4A. If the material is metallic, an inductive series detector is used; if the material is nonmetallic, a capacitive sensor is required. In case of a web break, a relay contact is made to sound an alarm or turn off a machine.

In figure 6–4b, a contactless cam switch can be made using the inductive detectors. Since the field strength and repeatability of the switch point is accurately controlled, the sensor can easily be made to detect only the high point on the cam and drop out on the low portion of the cam.

The Capacitive Proximity Detector

When we discussed the capacitive sensor element, we said that detection was made by the capacitor undergoing a capacitance change owing to the proximity of material. Detection is made by material approaching the sensor. The sensor's ability to detect is dependent on the dielectric contant of the target material. Each sensor is constructed to have a range of dielectric materials that it will detect.

Capacitive Sensor Operating Points   The operating point of the capacitive proximity sensor is necessarily different for all materials and, of course, all sensors. Each manufacturer of sensors provides a curve that describes how much of the sensors must be covered to produce an output signal. An example of this curve is illustrated in figure 6–5. Let's consider, for example, a steel target whose sensing distance is 8 millimeters. In order for the sensor to begin operation, 16 millimeters of the sensor would have to be covered. Each material has a

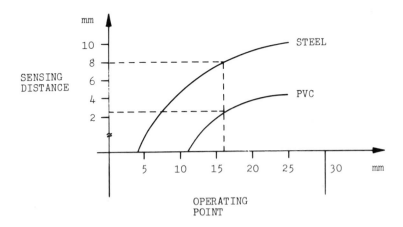

Figure 6–5   Operating Points Versus Sensing Distance of a Capacitive Proximity Detector *(Courtesy, Rechner Electronics Industries, Inc.)*

material scale factor. These scale factors are determined by the sensor manufacturers. Scale factors for this set of curves are as follows:

Material	Scale Factor
Steel or water	1.0
Wood	0.8
Glass	0.7
PVC (plastic)	0.4

To determine sensing distance for other materials using the same sensor, simply multiply the sensing distance for steel (or water) times the scale factor.

Example: 8 mm × 0.4 = 3.2 mm

If PVC covered 16 millimeters of the sensor it would be detected at 3.2 millimeters.

A Typical Capacitive Proximity Detector

Typical of the capacitive detectors is the Rechner KAS 70. The sensor is illustrated in figure 6-6. This sensor operates on dc voltage 10/30 vdc. Up to six of the sensors can be wired in series or parallel to perform logic functions.

Figure 6-6 A Capacitive Proximity Detector (*Courtesy, Rechner Electronics Industries, Inc.*)

Capacitive Proximity Detector Applications

The capacitive proximity detector is an extremely versatile device in that it is capable of detecting all materials, liquid and solid. Some of the applications are illustrated in figure 6-7.

Two types of sensor are available, flush-mount and non–flush-mount. Flush-mount sensors can be mounted side by side to cover a large area. If a powder or liquid is to be detected, a non–flush-mount sensor should be mounted inside the tank or bin. If a block of solid materials is to be detected, a flush-mount sensor should be used. Sensor wiring should be kept away from

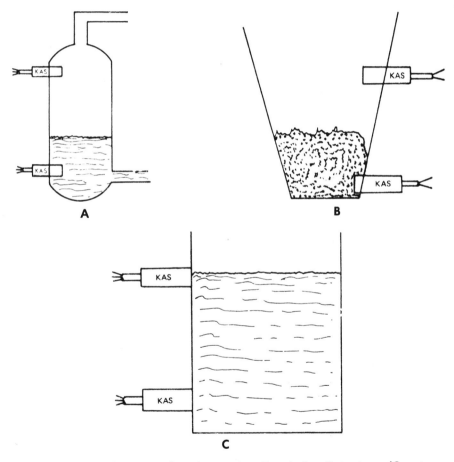

Figure 6-7 Applications for Capacitive Proximity Detectors *(Courtesy, Rechner Electronics Industries, Inc.)*

high-voltage lines or lines carrying inductive loads to prevent possible damage from high-voltage spikes.

A low- and high-level detection system can be made to work directly through the walls of nonmetallic containers (see figure 6–7A). The capacitive detectors will "see" through glass, fiberglass, wood, and plastic materials. Adjustable sensitivity is set to allow the exact level of liquid or solids to be detected during a manufacturing process. This method allows mounting without the necessity of liquid-tight joints. This same method can be used to detect water level in a boiler sight glass without pressure-tight joints. This advantage of sensing through the container can produce a reliable liquid-fill detector on production lines.

A level detection system for metal tanks can be made, using the capacitive sensors, by mounting the sensors through the walls of the tank. The sensor is held in place by two nuts on the threaded body and is made liquid-tight with rubber gaskets (see figure 6–7B).

This application can be particularly valuable to breweries and distilleries when the level of grains in storage bins must be known.

Small amounts of liquid that cling to the sensors when the liquid recedes will be ignored by the sensor. Only true level readings will cause the output to switch, thus eliminating false triggering, which can be damaging to the liquid pumping system.

A further use of these sensors is to detect powder or granular material in metal hoppers (see figure 6–7C)

The Magnetic Proximity Detector

A coil situated in a magnetic field will have a current induced in it if the magnetic flux changes. The magnitude of the induced current will depend on the rate at which the flux is changed. These are the basic principles on which the magnetic proximity detectors operate.

In its simplest form, a coil is wound around a bar magnet and one pole of the magnet is then located close to a ferrous object. If the ferrous object moves, the flux in the magnet changes and a current is induced in the coil. If a number of ferrous objects move past the magnet, a train of current pulses is induced in the coil, each pulse corresponding to the passage of one object past the magnet pole.

Magnetic pickoffs are most commonly used in conjunction with mild steel gearwheels, each tooth in the wheels being, in effect, a ferrous object. The pickoff is located radially and close to the periphery of the wheel and provides an electrical output having a frequency equal to the frequency of passage of the teeth past the pickoff.

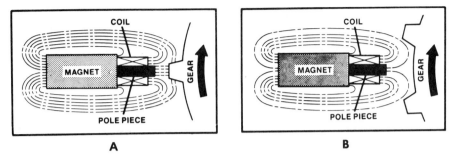

Figure 6-8 Low and High Reluctance Configurations for Magnetic Proximity Detectors *(Courtesy, Electro Corp.)*

Figure 6-8A shows a configuration setup for a low reluctance position. Figure 6-8B shows a configuration setup for a high reluctance position.

Factors Affecting Magnetic Proximity Sensor Output   The speed at which the magnetic flux is interrupted is directly proportional to the speed of passage of ferrous objects past the pickoff pole piece, but is attenuated by coil and iron losses which predominate at high frequencies. The output of magnetic pickoffs continues to rise with speed.

The output amplitude is inversely proportional to this distance separating the transducer pole piece and the moving ferrous object. In general terms, the output is greater for a greater mass of moving iron.

The design of modern magnetic pickoffs is a compromise to achieve a high-output voltage over a wide speed range in a small physical size. It is necessary that the active pole of the pickoff be small enough so that the sensing area is reduced to a level at which the transducer may be used effectively with small gearwheels. Load impedance is a final consideration.

Selection of a Magnetic Proximity Detector   The required output voltage will be determined by the sensitivity of the instrument receiving the sensor's signal. The first step is to determine the minimum operating speed at which a control signal is required. To convert rpm to surface speed of actuator in inches/second, use the following formula:

$$\text{ips} = \frac{\text{rpm} \times \text{gear diameter (inches)} \times \pi}{60}$$

where ips = inches per second

rpm = revolutions per minute

The second step is to determine the pole-piece clearance range, including actuator runout and tolerance stackup. The largest gap of this range should be used to determine what percent of standard voltage is available. Pole-piece clearance range is found in curves provided by the manufacturer of the sensor.

Gear diametral pitch may be determined by the following definition:

$$\text{pitch} = \frac{\text{number of teeth} + 2}{\text{diameter (inches)}}$$

Generally, the load impedance should be at least ten times the resistance of the sensor coil. For low-impedance loads, use the standard voltage-divider calculation.

It must be remembered that for every magnetic sensor there is an optimum gear-tooth configuration for maximum output. The illustration in figure 6–9 is an aid in determining typical relationships.

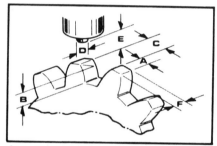

Gear Tooth Configuration.

Figure 6–9   Gear-Tooth Proximity Configuration *(Courtesy, Electro Corp.)*

In the illustration, optimum dimensions of A, B, and C are given as they relate to D, the diameter of the pole piece of the magnetic sensor being used. The optimum relationship for maximum output is as follows:

A—equal to or greater than D
B—equal to or greater than C
C—equal to or greater than three times D
E—as close as possible; typically .005 inch or less
F—equal to or greater than D

A Typical Magnetic Proximity Detector

The Electro Model 3070 is typical of the magnetic proximity transducers. The Model 3070 is illustrated in figure 6–10.

Figure 6-10 A Magnetic Proximity Detector *(Courtesy, Electro Corp.)*

Magnetic Proximity Detector Applications

The utilization of magnetic sensors usually falls into one of two categories. These are either *tachometry* (speed measurement) or *synchronization* (timing).

In tachometry applications, the magnetic sensor produces an output frequency, usually from an actuating gear, in direct proportion to rotational speed. The signal generated by the sensor in this mode is completely error-free and can be calculated for any given speed by the formula:

$$\text{frequency (f)} = \frac{\text{No. of gear teeth} \times \text{rpm}}{60}$$

The frequency thus generated can be converted directly to rpm by means of a frequency counter or digital tachometer. Another method of speed measurement is to change the frequency into a proportional dc current.

Examples of typical tachometer applications are: engine/motor/pump rpm sensing, over- or underspeed sensing, closed-loop speed-control feedback, wheel-speed detection, flow metering, transmission-speed sensing, tape and disc-drive rpm sensing, and synchronization of multiple-engine motor speed.

The zero-crossover point of the sensor signal repesents a highly accurate position reference, and it is this feature that makes magnetic sensors ideal for timing applications. The place where the output waveform changes from positive to negative is the "zero-crossover" point and corresponds to the time when the centerline of the pole piece and the centerline of the actuator are precisely aligned.

POWER-MEASURING TRANSDUCERS

The problems of measuring power are considerable. The variables involved are complex and numerous. Power variables have to do with voltage, current, resistance, and frequency. Each of these measurable components of electricity is also complex. Each has terms which, when monitored, may provide meaningful information to an engineer or a user. This part of the chapter is dedicated to several transducers that monitor power and/or its terms.

DIRECT CURRENT (DC)

Direct current is electrical current that flows in one direction. This sounds basic. However, the theories concerning the flow of current are numerous and extensive, and some are far from basic. With study, though, all theories fall into place.

The common characteristics of all types of electrical currents are their variations in amplitude, time, and direction. Direct current was the first current to be widely used and understood. Its applications are relatively simple. However, it has some serious limitations, among which are the facts that it does not develop waveforms and cannot be transmitted (radiated) by an antenna.

If direct current (dc) as we know it today is transmitted by wire for any great distance, the power loss because of wire resistance tends to be nearly as great as the power originally transmitted. Therefore, direct current is impractical to transmit except for a very short distance. It is thus most useful for fixed-position jobs such as in power machinery, driving mechanisms, and portable machinery that has no immediate access to a supply of alternating current. Direct current is also used extensively for relay controls and solenoid switching.

Probably the best use of dc electrical power is for electronic circuitry, where the direct current provides a "platform" for the ac signals to ride upon or develop around. All electronic circuitry has some amplitude of dc with which the circuit is biased. This level of dc is increased or decreased by varying ac or dc signals.

The purity of dc power is always questionable. No absolutely pure, smooth, direct-current source is available; it all varies somewhat. Designers go to great pains to make dc power supplies that do not vary in amplitude, to ensure that signal development around the dc levels are not affected by changing dc levels. To a practical extent they succeed. Perhaps, then, the greatest advantages of dc power are its characteristics of constancy in amplitude, time, and direction.

FORMULAS FOR POWER IN DIRECT-CURRENT CIRCUITS

The formulas for direct-current power terms are centered around Ohm's law. Power terms are *voltage, current, resistance,* and *power.*

$$E = IR$$
$$I = \frac{E}{R}$$
$$R = \frac{E}{I}$$

where E = voltage in volts (V)

I = current flow in amperes (A)

R = resistance in ohms (Ω)

Power formulas are somewhat different but still utilize the same terms.

$$P = EI$$
$$P = I^2R$$
$$P = \frac{E^2}{R}$$

ALTERNATING CURRENT (AC)

Alternating current is electrical current that varies in magnitude and reverses in polarity. A pure sine wave of alternating current continuously varies in magnitude and periodically reverses direction. The electric charge (electrons) moves back and forth along the circuit like a weight on a string (a pendulum) swings. Such motion is sinusoidal or simple harmonic.

Alternating current is much more versatile than direct current. The basic fact is that it is easily transferable without much power loss. Direct current loses much power in transfer.

Alternating current is able to radiate at different time intervals (frequencies). This makes it useful in most electronic fields such as radio, radar, television, and so forth.

The parts of a sine wave are important to measurement. These parts of a sine wave are easily shown with the diagram in figure 6–11. One complete cycle is achieved in 360 degrees of rotation, as shown in the development of a sine wave. A single alternation (half of a cycle) is completed at 180 degrees. Maximum positive amplitude to maximum negative amplitude is peak to peak (voltage or current). Maximum amplitude to minimum amplitude (zero) is peak (voltage or current). Average current or voltage is 0.636 × peak and effective current or voltage (rms) is 0.707 × peak.

An instantaneous voltage is at any point on the sine wave and is the voltage existing at any one instant of time. Average voltage is the average of all the points of an alternation. This average is taken from a single alternation, because the average of both alternations would equal zero. Effective voltage is

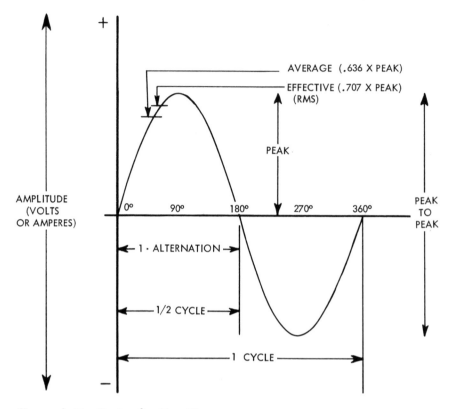

Figure 6-11   Parts of a Sine Wave

equal to the trigonometric value of the sine of 45 degrees. It is also called rms or root-mean-square value (used in commercial power-line calculations).

Conversion from one value of ac measurement to another is a simple matter of multiplication. Conversion is often necessary in maintenance and trouble-shooting processes. Table 6–1 illustrates the measurement and how to convert it to some other ac measurement.

TABLE 6-1. AC Measurement Conversion (Sine Waves)

Conversion From	Multiplication Factor			
	RMS	Average	Peak	Peak to Peak
RMS (Effective)	1.00	0.900	1.414	2.828
Average	1.110	1.000	1.570	3.141
Peak	0.707	0.636	1.000	2.00
Peak to Peak	0.354	0.318	0.500	1.000

FORMULAS FOR POWER IN ALTERNATING-CURRENT CIRCUITS

There are many power formulas for ac circuits. These formulas are listed later in this section. There are some points concerning power, however, that should be cleared up before we concern ourselves with formulas.

Real power can be calculated using the total resistance in the circuit. Real power is measured in watts. The formula $P = I^2R$ is used. In the event voltage and current are used in a circuit with reactance and resistance, real power is calculated using the formula $P = EI \cos < \theta$. This is required because of the phase angle between current and voltage. Either formula can be used depending on which is more suitable for the situation.

The power factor in a series ac circuit is the ratio of resistance to impedance, or $PF = \dfrac{R}{Z}$. In a parallel circuit the power factor is the ratio of resistive current to total current, or $PF = \dfrac{I_R}{I_T}$.

Apparent power is that the power in a circuit which has no resistance. It is calculated by multiplying voltage times current. Since current and voltage are 90 degrees out of phase in pure reactance, power is only apparent. Real power is measured in watts (W). Since no power is actually dissipated in pure reactance, the formula $P = EI$ is measured in volt-amperes (VA) when one is speaking of reactive circuits.

The United States still uses the customary U.S. system of measurements for most technologies except electricity, where the units are metric, such as amperes, volts, ohms, and watts. But in some applications, particularly in power electricity, electrical units have to be translated or converted to mechanical units such as horsepower (HP) in the customary U.S. system. This is possible because 746 watts or 0.746 kilowatt (real power) are equivalent to one horsepower. For the benefit of users, electric motors in the United States are therefore rated in horsepower or fractional horsepower.

In the SI (International System) of metrics proposed for adoption by the United States, mechanical power, as well as all other forms of power, is measured in watts (W) or kilowatts (kW). Knowing this does not, however, make it an accomplished fact at this time.

Formulas for Voltage (E) in Sinusoidal Alternating Current Circuits

$$E = IZ$$
$$E = \frac{P}{I \cos \theta}$$
$$E = \sqrt{\frac{PZ}{\cos \theta}}$$

Formulas for Current (I) in Sinusoidal Alternating Current Circuits

$$I = \frac{E}{Z}$$

$$I = \frac{P}{E \cos \theta}$$

$$I = \sqrt{\frac{P}{Z \cos \theta}}$$

Formulas for Impedance (Z) in Sinusoidal Alternating Current Circuits

$$Z = \frac{E}{I}$$

$$Z = \frac{P}{I^2 \cos \theta}$$

$$Z = \frac{E^2 \cos \theta}{P}$$

Formulas for Power (P) in Sinusoidal Alternating Current Circuits

$$P = IE \cos \theta$$

$$P = I^2 Z \cos \theta$$

$$P = \frac{E^2 \cos \theta}{Z}$$

THE HALL EFFECT TRANSDUCER (See figure 6-12)

The Hall effect was discovered by E. H. Hall in 1879. The Hall effect, for purposes of this discussion, is that characteristic of a certain crystal such that when
it is conducting current (control current) and is placed in a magnetic field, a
potential difference is produced across its opposite edges. The potential difference is proportional to the product of the control current, the strength of the
field, and the cosine of the phase angle between the control current and
magnetic flux. By putting the crystal in a magnetic structure such that an ac current generates the flux and an ac voltage produces the control current, we have a
multiplying device. It produces an output proportional to IE cos θ (power).
This is in essence the Hall generator watt transducer. In figure 6-12, the load
current I_{ac} produces a proportional flux through the crystal in the air gap. The
load voltage E produces a proportional control current through the crystal. The
output is proportional to the product of the two and the phase angle between
them.

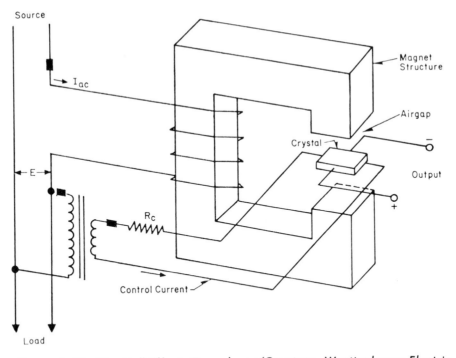

Figure 6-12   The Hall Effects Transducer *(Courtesy, Westinghouse Electric Corp.)*

Actually, the output of the Hall generator consists of a dc voltage proportional to true power (watts) plus a double-frequency ac voltage proportional to the volt-amperes in the circuit. When the transducer is used with devices that do respond to ac, the double-frequency component must be filtered out. Filters are provided especially for this purpose.

The output of a Hall element decreases with increasing temperature. The watt transducers are temperature-compensated with a thermistor-resistor network in the output circuit. The load resistance is part of the network; therefore it is fixed for a particular transducer.

The Hall Watt Transducer (see figures 6-13, 6-14, and 6-15)

The Hall watt transducer is constructed in three basic types: the *single-phase transducer*, the *three-phase three-wire transducer*, and the *three-phase four-wire transducer*.

The single-phase, one-current-coil (1-element) watt transducer is il-

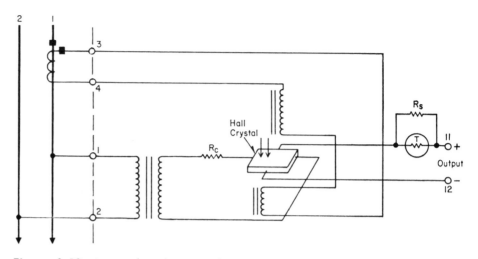

Figure 6–13   Internal and External Wiring of a Single-Phase Watt Transducer *(Courtesy, Westinghouse Electric Corp.)*

lustrated in figure 6–13. This transducer has one current coil and one potential coil. It measures watts in single-phase, two-wire circuits.

In applying the Hall principles to an ac wattmeter, a Hall crystal is mounted in the air gap of an electromagnet with a current circuit supplying the energy to the electromagnet and a potential circuit connected to two opposite sides of the Hall crystal. Output of the element is derived from the two sides of the crystal at right angles to the potential input. The output is pulsating dc, which is measured by a conventional permanent-magnet, moving-coil instrument mechanism. This output may be used with other devices for either measurement or control, within the specified ratings.

The watt transducer operates on the Hall effect principle, producing a dc output which is proportional to ac power input. Referring to figure 6–13, in the single-phase transducer, line voltage is connected across terminals 1 and 2, causing current I_c to flow through calibrating resistor R_c and the Hall crystal. Line current is connected to terminals 3 and 4, producing a proportional flux perpendicular to the plane of the crystal. The output voltage of the Hall crystal (E_o), appearing across terminals 11 and 12, is proportional to the current through the crystal, the flux perpendicular to it, and the cosine of the phase angle between them. Expressed as a formula, $E_o = K_1 I_c \phi \cos \theta$, where K_1 is the Hall crystal proportionality constant, I_c is the current through the crystal, ϕ is the flux produced by the electromagnet circuit, and $\cos \theta$ is the phase angle between this current and the flux.

In figure 6–14, the internal and external wiring of a three-phase, three-wire Hall watt transducer is illustrated. This is a polyphase, two-current-coil

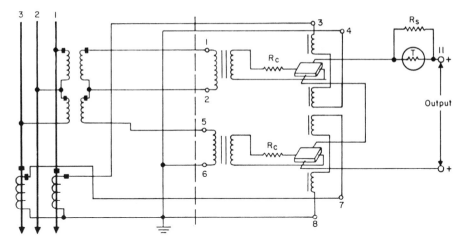

Figure 6–14 Internal and External Wiring of a Three-Phase, Three-Wire Watt Transducer *(Courtesy, Westinghouse Electric Corp.)*

(two-element) circuit. There are two current coils and two potential coils for the measurement of three-phase, three-wire power. They are also used on single-phase, three-wire and on certain three-phase, four-wire applications (where the potential transformers are connected line to line). The outputs of the two Hall elements are connected in series.

In figure 6–15, the internal and external wiring of a three-phase, four-wire

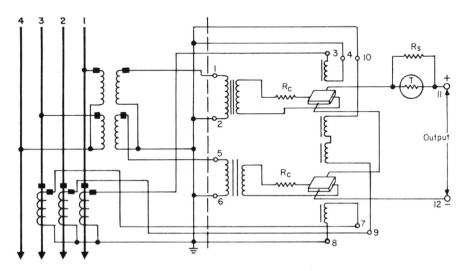

Figure 6–15 Internal and External Wiring of a Three-Phase, Four-Wire Watt Transducer *(Courtesy, Westinghouse Electric Corp.)*

watt transducer is shown. This is a polyphase, three-current-coil (two-and-a-half-element) circuit. This transducer has three current and two potential coils. The two Hall elements each have two current coils, with one coil in each element being connected in series to form the third current coil. This transducer, theoretically, is accurate on balanced voltages only.

The Hall VAR Transducer

Watts or true power, as measured by a Hall watt transducer, are the product of in-phase voltage and current. VARs or reactive power is the product of voltage and that portion of current which is 90° out of phase with the voltage. This phase shift is accomplished with a polyphase phase-shifting transformer. Essentially, watt transducers and VAR transducers are the same.

CURRENT TRANSDUCER (See figure 6-16)

The transducer converts an ac current input to a proportional dc current of low magnitude. It consists of a current transformer with a loading resistor feeding through a calibrating rheostat to a full-wave rectifier. There is an R-C network in the circuit for waveform error compensation to compensate for errors that result from moderate waveform distortion. (This compensation method is known as 120° commutation.)

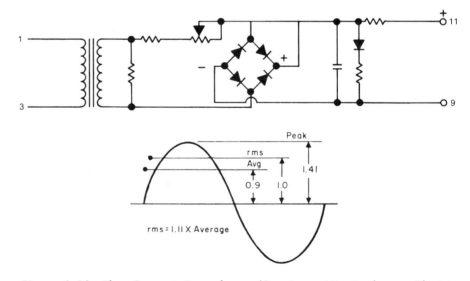

Figure 6-16 The Current Transducer *(Courtesy, Westinghouse Electric Corp.)*

The perfect sine-wave alternating current has the relationship shown in figure 6–16. Electrical power measurements are made in terms of the rms values which yield true power or "heating effect" of a current. These values are measured directly by the dynamometer or iron-vane mechanisms used in conventional ac instruments. The rectifier-type instruments used in test sets and in lower-cost switchboard instruments respond to the average values only. Thus, rectifier instruments read rms values only when they are used on perfect sine waves.

On power systems, a pure sine wave is rare. Close to the source, the voltage wave tends to be clean but the current wave is distorted. The opposite is true of the far end of a system, generally. Distortions of waveform are caused by harmonics generated in equipment, partial wave utilization of controlled firing rectifiers, capacitors, etc.

VOLTAGE TRANSDUCER

The voltage transducer is the same as the current transducer except for the input transformer and the omission of the transformer loading resistor.

THE FREQUENCY TRANSDUCER (See figure 6–17)

The frequency transducer provides a linear dc output on each side of a null point within the frequency span of the device. This makes the transducer essentially failsafe when used with an indicating instrument. An input voltage of 12 volts ac is applied between terminals 1 and 3. Resistor R1 limits the input

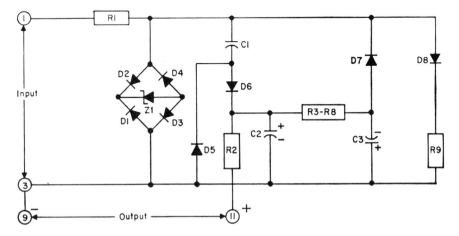

Figure 6–17 Frequency Transducer Schematic *(Courtesy, Westinghouse Electric Corp.)*

voltage which is then clipped by the Zener-diode bridge network (D1, D2, D3, D4, Z1) producing a near square wave of constant amplitude voltage at this point. Resistor R9 and diode D8 are used to balance the loading of the Zener circuit. On one half-cycle of the signal, capacitor C1 discharges through diode D6 and R2 (the output load calibrating resistor) against the Zener voltage.

The combination of diode D7, capacitor C3, and resistor R3-R8 produce a nulling current at the junction of diode D6 and resistor R2 which balances the circuit to provide zero output.

The integrated discharge current of capacitor C1 changes with input frequency ($i = 2$ fce), causing a proportional change in the output. Capacitor C2 is used to filter out part of the ac ripple in the output. The transducer is calibrated for specific load resistances.

The transducer produces an output that is load-sensitive beyond a 10% variation from the calibrated load.

FILTERING (See figure 6–18)

Filters are for use with transducers with dc outputs having unwanted ac ripple. The output indicating instrument must be of high impedance in respect to the input impedance. The output instrument should have 50,000 ohms minimum

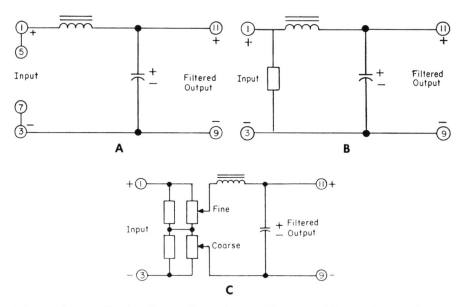

Figure 6–18 Filtering-Power Transducers *(Courtesy, Westinghouse Electric Corp.)*

for a 50-ohm input load. A typical application is a null balance potentiometer or A/D converter.

The filter as shown in figure 6–18A contains an inductor and capacitor filter network with provisions for connecting an external input resistor.

The filter as shown in figure 6–18B contains an additional 50-ohm $\pm$ $\frac{1}{2}$-watt input resistor.

The filter in figure 6–18C contains two potentiometers for fine and coarse (0 to 100 percent) adjustment of the voltage output. The input resistance is 50 ohms $\pm$ $\frac{1}{2}$ percent.

REFERENCES

Reference data and illustrations were supplied by state-of-the-art manufacturers. Permission to reprint was given by the following companies:

Inductive and capacitive proximity sensors—Rechner Electronics Industries, Niagara Falls, New York

Magnetic proximity detectors—Electro Corporation, Sarasota, Florida

Power-measuring transducers—Westinghouse Electric Corp., Pittsburgh, Pennsylvania and Coral Springs, Florida

All copyrights © are reserved.

7

Fluids: Flow, Level, and Pressure

One of the broadest areas in the transducer field is that of fluid transducers. When we speak of fluid parameters we must consider flow, levels, and pressure of both hydraulic and pneumatic fluid (liquid and gas). There are, of course, other parameters involved such as temperature and viscosity. The objective of this chapter is to provide the reader with some information about fluids at rest and in motion. Then we shall provide you with operation details and examples of typical fluid transducers used today. Specific operation of these transducers will be explained.

FLUIDS AT REST

Fluid pressure is defined as the force per unit exerted by a gas or a liquid on the walls of a container. Pressure is the prime measure of the fluid's energy content and therefore a key engineering parameter. It is imperative that we understand the nature of pressure and the relationship to other parameters.

Pressure in Liquids

Unlike solid substances, liquids and gases are fluids that completely lack rigidity, moving about freely and conforming to the shape of their container. Whenever a body of fluid is at rest, all forces acting on it or within it must be completely in balance, otherwise the fluid will flow until equilibrium is established and rest is achieved.

For example, in a stationary pail full of water, picture an imaginary column of water measuring 1 inch square in cross-section and extending downward from the surface a distance of H inches (see figure 7-1). Since the li-

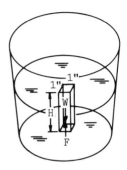

Figure 7-1   Weight of a Square Column of Water
(Courtesy, Computer Instruments, Corp.)

quid is at rest, the entire weight of this column (W) must be supported by an equal upward force (F) exerted by the liquid outside the column on the bottom surface of the column. The weight of the column acting downward is equal to the volume of water multiplied by its density. The density of water is nominally 62.4 pounds per cubic foot or .036 pound per cubic inch. Thus the weight of the column is 1 inch × 1 inch × H × .036 = .036H pound. The upward-acting supporting force distributed over the 1-inch-square bottom of the column is therefore also .036H pound. Since pressure is defined as the force per unit area, a 10-inch column of water weighing .36 pound exerts a pressure of .36 pound per square inch (psi). This is the pressure of the liquid at the 10-inch depth. Since in the liquid at rest all forces must be in balance everywhere in it, the pressure at any point in the liquid at a given depth is the same in all directions.

For a given liquid at rest, the internal pressure at any point within the liquid is directly proportional to the liquid height above the point. The more general relationship states that the pressure is the product of liquid height, nominal density of water, and liquid specific gravity relative to water:

$$P = H \text{ inches} \times .036 \times G \text{ psi}$$

where P = the internal pressure in psi

H = height in inches below the free surface of a liquid at rest

G = specific gravity of the liquid

The specific gravities of various common liquids are listed below:

Liquid	Specific Gravity	Temperature
Water, pure	1.000	(4°C)
Water, sea	1.025	(15°C)
Alcohol, ethyl	.079	(0°C)
Glycerine	1.26	(0°C)
Mercury	13.596	(0°C)
Oils, lubricating	0.90-0.93	(20°C)

This pressure, sometimes called hydrostatic pressure, is independent of the size of the container or its shape. At the bottom of a 10-inch-deep pond or a 10-inch test tube filled with water, the pressure is the same—.36 psi; at half the depth, 5 inches beneath the surface, the pressure is .18 psi. The pressure at the bottom of a 10-inch test tube filled with mercury (13.6 times as dense as water, G = 13.6) is $13.6 \times .36 = 4.9$ psi; this is the same as the pressure at the bottom of a water pond 136 inches deep.

Pressure can be equivalently expressed in terms of pounds per square inch or per square foot, in inches or feet of water or any other reference liquid such as mercury, or in their metric equivalents.

Pressure in Gases

Although classified as fluids because they flow and adapt themselves to the shape of the containing vessel, gases differ from liquids in two respects: (1) gases are very compressible, and (2) they expand to fill any closed vessel in which they are placed. Otherwise they behave as liquids do. So, just as the weight of liquid in a pail or in the ocean exerts a pressure at any depth below its free surface, the ocean of air surrounding the earth acts on its surface. This body of air, like all fluids, exerts a pressure determined by its height and density.

The net effect of the total weight of the atmosphere is to exert a mean pressure of 14.7 psi (equivalent to about 34 feet of water or about 30 inches of mercury) on the earth's surface at sea level. The familiar barometric reading in weather reports refers to this pressure.

Mean atmospheric pressure is sometimes used as a term of reference. Particularly when dealing with high gas pressures, these may be expressed as a multiple of atmospheres; for example, a pressure of 2,940 psi is sometimes spoken of as 200 atmospheres (2940/14.7).

On account of its compressibility, the air near the earth is weighted down and compressed by that above, and as a result is more dense than at higher elevations. Consequently, the pressure does not vary uniformly with altitude, as in a medium of uniform density like a liquid, but changes less and less rapidly at greater heights.

Almost all substances expand with increasing temperature, becoming less dense. Typically, the density of water is expressed at a given temperature. To account for gas compressibility, it is necessary to state density at a given pressure as well as at a given temperature. So at standard temperature and pressure of 0°C and 30 inches of mercury, the density of air is 0.081 pound per cubic foot; water is about 770 times as dense. Because of this great difference in densities, the specific gravity of gases is usually expressed with reference to air as a standard rather than water. The densities and specific gravities of a few common gases are given:

Gas	Density lb/cu. ft.	Specific Gravity
Air	0.081	1.000
Carbon Dioxide	0.123	1.529
Hydrogen	0.0056	0.069
Helium	0.011	0.138
Nitrogen	0.078	0.967
Oxygen	0.089	1.105
Steam at 100°C	0.037	0.462

There is a basic relationship between pressure, volume, and temperature of a gas in a closed container expressed by the general gas law:

$$\frac{PV}{T} = \text{constant}$$

where P = absolute pressure (above a vacuum)

V = gas volume at the temperature (C)

T = absolute temperature (273 + t)

If a given body of gas is compressed to a smaller volume, its internal pressure will increase; for example, the pressure of a fixed quantity of trapped air in a bicycle tire pump increases as the piston movement causes the volume to decrease. Similarly, the pressure of a body of gas in a closed container will rise when the gas is heated; the home pressure cooker uses this effect.

FLUIDS IN MOTION

The pressures existing within fluids at rest were previously discussed. They were referred to as liquid hydrostatic pressure or gas static pressure. This static pressure or head has the potential for doing useful work—i.e., moving the fluid from one place to another. For example, the static water head in an elevated city reservoir ultimately is translated into water gushing out of the home sink tap.

Pressure and Flow

It has been previously discussed under the section "Fluids at Rest" that the static pressure at any level of the liquid in a tank can be expressed as a quantity equal to the linear depth of the level below the free surface. Referring to figure

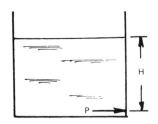

Figure 7-2 Hydrostatic Pressure *(Courtesy, Computer Instruments, Corp.)*

7-2, the pressure against the outside wall of a water tank at a depth H inches below the free level is equal to a pressure of .036 × H″ pounds per square inch (psi), equivalent to a pressure of H inches of water. This pressure is called the static head. The work potential (or stored potential energy) is equal to the weight of the fluid multiplied by its elevation:

$$\text{Potential energy} = W \times H$$

$$\text{where } W = \text{weight of the fluid}$$

$$H = \text{depth in inches}$$

A fluid in motion is a moving mass and possesses kinetic energy. The kinetic energy is equal to one-half its weight multiplied by the square of its velocity v and divided by the gravity constant g:

$$\text{Kinetic energy} = \frac{Wv^2}{2g}$$

$$\text{where } v = \text{velocity of the fluid}$$

$$g = \text{gravity constant}$$

If the fluid in the tank were suddenly released, say through a hole in the tank's bottom or side, then the potential energy would be converted to kinetic energy. If all the potential energy is converted without loss into fluid motion (an ideal case, assuming a frictionless fluid), then the kinetic energy equals the potential energy:

$$WH = \frac{Wv^2}{2g}$$

$$\text{or} \quad H = \frac{v^2}{2g}$$

$$\text{where} \quad H = \text{the } static\ head$$

$$\text{and the term} \frac{v^2}{2g} = \text{the } velocity\ head$$

The equation for static head expresses a basic relationship between pressure and flow. In various statements it is used to describe the operation of many flow-measuring devices.

The velocity of flow resulting from a static pressure head H may be found with the formula:

$$v = \sqrt{2gH}$$

When H is in feet and the gravity constant is taken as 32.2 feet per second, then the velocity in feet per second is:

$$v = 8.02 \ \sqrt{H} \ \text{ft/sec}$$

Orifices, Nozzles, and Weirs

In figure 7-3, the potential energy of the stored liquid is converted to kinetic energy in the liquid exiting from a tap. Liquid exits from tank orifices at a velocity proportional to $\sqrt{H}$ where H is the static head in feet. In accordance with the equation $v = 8.02 \ \sqrt{H}$ feet/seconds, the exit velocity is proportional to the square root of the height of the liquid surface above the orifice or nozzle opening, as shown.

Assuming an ideal no-loss discharge through the submerged orifice, the volume Q of liquid discharged per unit time is

$$Q = A \times v$$
$$\text{or } Q = A \sqrt{2gH}$$

where A = the cross-sectional area of the orifice opening

H = the height of the liquid surface above an orifice

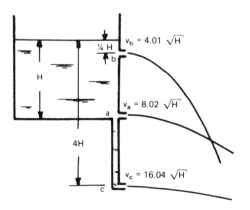

Figure 7-3 Orifice or Nozzle Flow *(Courtesy, Computer Instruments, Corp.)*

Practically, however, losses do occur in the discharge through an orifice or a nozzle owing to fluid friction, turbulence, and the like. The volume of liquid actually discharged, somewhat less than ideal, is

$$Q = CA\sqrt{2gH}$$

where C = the orifice or nozzle discharge coefficient ranging in value from 0.60 for a sharp-edged circular orifice to about 0.97 for a smooth rounded-approach one

The *weir* is a form of orifice used in open channels in which the water passes over the sharp upper edge of a submerged vertical plate spanning the channel. In an open channel, the submerged weir spans the channel and forms a sharp-edged orifice past which the liquid flows at a velocity proportional to H, the upstream static head (see figure 7-4). The static head producing flow is the water level H above the weir plate, measured upstream from it. The effective cross-sectional "area" of this type of rectangular orifice is the weir width L multiplied by 2/3 H. Inserting this area term into the equation, the volume of liquid discharged per unit time over a weir becomes:

$$Q = C2/3HL\sqrt{2gH}$$

where C = the discharge coefficient, typically 0.62 for a deep sharp-crested weir

The relationship between flow and static head, expressed in the last two equations, makes possible the direct reading of fluid flow by measurement of the static pressure head.

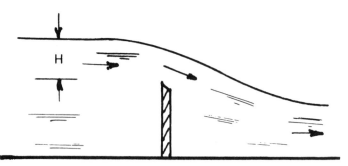

Figure 7-4 Flow across a Weir *(Courtesy, Computer Instruments, Corp.)*

A gage pressure transmitter tapped into the tank or open channel wall, with proper zero offset to compensate for vertical tap location relative to orifice or weir crest, provides an electrical signal proportional to the static head H. Feeding this signal to electronic circuitry suitable for solving the last two equations, sometimes called a "square-root extractor," converts it into an output proportional to fluid volume discharge rate.

Pitot Tubes

If the motion of a liquid having a velocity head $v^2/2g$ is arrested, then the velocity head representing its kinetic energy is converted into a static head equal to the equivalent potential energy. When the current in a moving open stream is directed against the small opening of the Pitot tube (see figure 7–5A), the liquid in its stem rises to a height above the surface of the stream equal to the velocity head. A gage pressure transmitter connected to the Pitot tube will provide an electrical signal proportional to this head, and by the use of circuitry for the solution of the equation $v = \sqrt{2gH}$, will provide an electrical signal proportional to velocity (see figure 7–5B).

When the liquid is flowing under pressure in a pipe, arresting its motion in a Pitot tube (see figure 7–6A) will produce a height of liquid in its stem equal to the total pressure consisting of the sum of the static head and the velocity head; the height of liquid in the tube stem of a second tap through the pipe wall, which does not enter the stream, will rise only to the static head level. A differential pressure transmitter connected between the Pitot and static pressure tubes (see

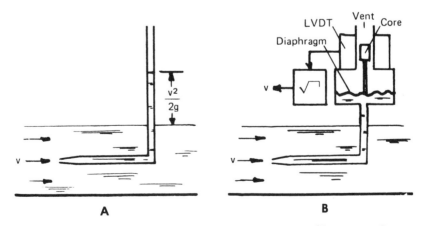

Figure 7–5 Fluid Flowing in a Moving Open Stream *(Courtesy, Computer Instruments, Corp.)*

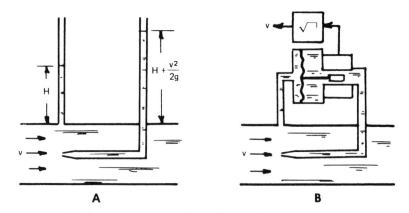

Figure 7-6 Fluid Flowing in a Closed Pipe *(Courtesy, Computer Instruments, Corp.)*

figure 7–6B) will provide a signal proportional only to the velocity head or, by suitable circuitry, to the velocity itself.

Venturi Meters

When a volume of liquid in a pipe flows under pressure from one place to another (see figure 7–7), the total head (which is the sum of static and velocity heads) remains constant—that is, except for friction losses. In an ideal liquid, the sum of the pressure head is the same everywhere in the pipe. This is known as the Bernoulli Law:

$$H_a + \frac{v_a^2}{2g} = H_b + \frac{v_b^2}{2g}$$

To pass through a given volume of fluid, the velocity in a narrow section of pipe will be greater than in a wide section of pipe. The flow through a Venturi

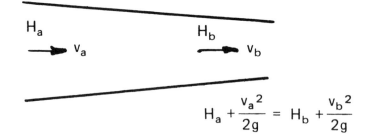

Figure 7-7 Bernoulli's Law *(Courtesy, Computer Instruments, Corp.)*

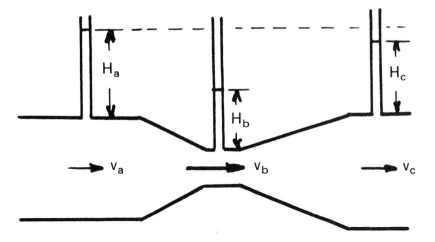

Figure 7-8 Flow through a Venturi Meter *(Courtesy, Computer Instruments, Corp.)*

meter (see figure 7–8), having an inlet pipe cross-sectional area (a) and throat cross-section area (b), will be at a greater velocity in the throat than in the inlet.

Since the velocity head in the throat will be greater than in the inlet, the static head will be less in the throat (refer to the last equation), as indicated by the relative liquid heights in the static pressure taps.

In such a device the volume of liquid discharged in a unit time is:

$$Q = CA \sqrt{\frac{2g\,(Ha - Hb)}{a^2/b^2 - 1}}$$

where in a properly proportioned Venturi the discharge coefficient (c) may be close to 1.

A differential pressure transmitter connected between the static pressure taps of a Venturi meter, with suitable electronic circuitry for solving this equation, will provide an electrical signal output proportional to volumetric flow (see figure 7–9)

Pipe Orifices

A simpler device than the Venturi meter for constricting the pipe cross-sectional area, but one less efficient, is the sharp-edged orifice shown in figure 7–10. A widely used means of instrumenting for flow monitoring is to measure the pressure difference between the upstream and downstream static pressure taps

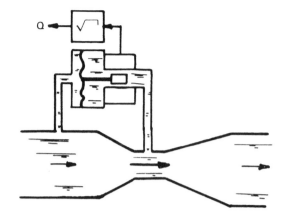

Figure 7-9   Flow through a Venturi Meter with a Differential Transmitter *(Courtesy, Computer Instruments, Corp.)*

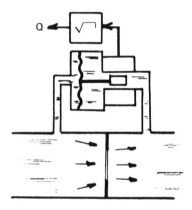

Figure 7-10   Flow in Pipe Orifices *(Courtesy, Computer Instruments, Corp.)*

with a differential pressure transmitter which includes the circuitry necessary for solving the last equation.

Flow in Bends

When water flows through a pipe bend, centrifugal force creates a difference between the pressures at the inside and outside of the pipe bend (see figure 7-11). In such a device, the volume flow through the bend is:

$$Q = CA \sqrt{2g(H_o - H_i)}$$

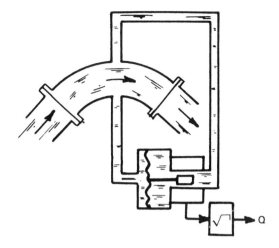

Figure 7–11 Flow in Bends *(Courtesy, Computer Instruments, Corp.)*

where the discharge coefficient C will have values between 0.56 and 0.88, depending upon the size and shape of the flow bend.

A differential pressure transmitter connected between the static pressure taps at the inside and outside of a pipe bend, with suitable electronic means for solving the aforementioned equation, will provide an electrical signal output proportional to volume flow.

Practical Flow Measurement

As a consequence of the basic energy relationships, the measurement of flow typically involves computing the square-root value of the pressure head or pressure-head difference. When the flow range is limited—i.e., the ratio of maximum to minimum flow rate is less than 5 or 6 to 1—the error in square-root computation will be a small multiple of the error in the pressure-transmitter measurement. In installations where the flow range is larger, the pressure-transmitter errors in the square-root extraction become greatly amplified and other techniques of pressure sensing must be used; for example, paralleled flow-measuring devices.

Gas Flow Versus Liquid Flow

When the flowing fluid is a gas rather than a liquid, account must be taken of the compressibility of the gas in properly proportioning nozzles, Venturi meters, and orifices, as well as locating the pressure taps. Otherwise, the static-pressure measurements will not be true indicators of flow because of local pressure anomalies. The discharge coefficients likewise may be in error unless calibration is performed with the specific gas in question. As a general rule, however, where flow constrictor devices are so sized that small pressure drops are encountered—i.e., where constriction is small—the flow equations for incompressible liquids can be applied to gases with small error.

Real Fluids Versus Ideal Fluids

The constancy of total head, expressed in the Bernoulli equation listed under the section "Venturi Meters," is true for an ideal, friction-free passage (or pipe). Real fluids absorb some portion of the static head in overcoming friction while moving from one place to another; in very small-diameter tubing, or for thick liquids, static head is required to overcome the effects of viscosity as well. Consequently, in figure 7–8, in an ideal fluid moving from inlet (a) to throat (b) and then to outlet (c), H_a and H_c would be equal, indicating 100 percent pressure recovery. In a real fluid, H_c would be less than H_a. The difference is accounted for by loss due to friction, turbulence, or other factors of the moving fluid and the pipe.

Summary of Fluids in Motion

Orifices, nozzles, Pitot tubes, Venturi meters, and pipe bends are fundamental devices for producing a regular and reproducible pressure difference that is related to rate of flow. Sensing fluid pressure in tanks, channels, pipes, and the like thus provides a basic means for measurement and control in a wide variety of practical situations.

FUNDAMENTALS OF GAS METERS

The 50 million gas meters currently in service with the different phases of the gas industry in the United States, plus the majority of a similar number of meters installed elsewhere in the world, use two different physical principles to measure gas volumes. These two physical principles are *positive displacement,* comprising the large majority, and *inferential meters,* used primarily for large-volume flows.

Positive Displacement Meters

Meters incorporating the positive displacement principle of measurement are of the diaphragm and rotary types.

In positive displacement measurement, a barrier of some sort is inserted in the gas stream to separate the unmetered upstream gas from the metered downstream gas. Precisely known volumes of gas are transported across this barrier during each cycle of the measuring device. Adjustments are employed to calibrate the volume per cycle to desired engineering units. The product of the volume trapped per cycle times the number of cycles is displayed on any of a wide variety of readout devices as totalized volume at line conditions.

Consider the formula:

$$Vm = Vc \times Nc$$
$$\text{where } Vm = \text{totalized volume}$$
$$Vc = \text{volume per cycle}$$
$$Nc = \text{number of cycles}$$

Inferential Meters

These design limitations for positive-displacement meters have led to the development of meters which utilize the second basic measuring principle, called *inferential measurement*.

The dictionary definition of the word "infer" is "to derive a conclusion from observed facts or evidence." Inferential gas meters follow this definition. The observed physical fact of velocity through a known area is used to derive volume measurements.

There are two basic types of inferential gas meters: *orifice meters* and *turbine meters*.

A turbine meter introduces a restriction (called a nose cone) of known cross-sectional area into the gas stream, as does an orifice meter (see figure 7-12). However, the turbine meter determines flow velocity through this restriction by counting rotations of a turbine rotor mounted in the open or

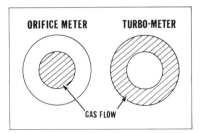

Figure 7-12 Gas Flow—Orifice versus Turbometers *(Courtesy, Rockwell International)*

"throat" area of the restriction. The turbine-blade rotations are transferred through a gear train to a wide variety of readout devices where totalized volume at line conditions is displayed.

Orifice Meters (See figure 7–13)   When a fluid flowing through a closed duct encounters a restriction, a local pressure drop is developed. The magnitude of the pressure drop is related to the flow rate at which the fluid flows through the duct.

Note that the pressure drop is related to flow rate, not to volume. Timing devices are used to record the time/flow rate data and the integration of these elements is used to produce volumetric readings. Thus, the orifice meter is basically a velocity meter and the volumetric measurement is derived or inferred from the observed fact of velocity versus time.

Turbine Meters (See figure 7–14)   Gas turbine meters are velocity-sensing devices, as are orifice meters. The direction of flow through the meter is parallel to a turbine rotor axis and the speed of rotation of the turbine rotor is nominally proportional to the rate of flow. Gas volumes are derived or "inferred" from the rotations of the turbine rotor.

The essential function of a gas meter is to provide an accurate readout of totalized volume throughout. It is in the fulfillment of this function that the turbine meter excels. Direct readouts can be displayed on any of a wide variety of

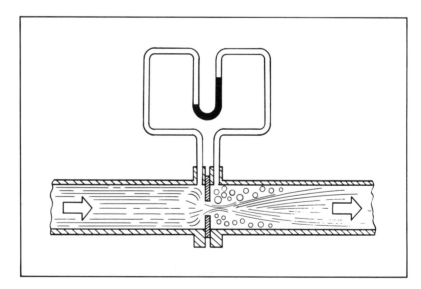

Figure 7–13   Orifice Meter Operation *(Courtesy, Rockwell International)*

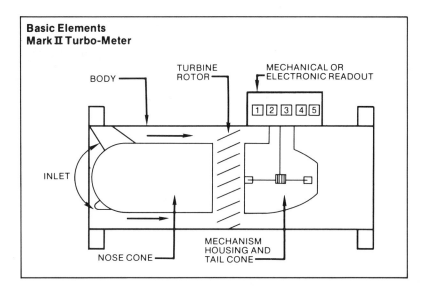

Figure 7–14 Gas Turbine Meter *(Courtesy, Rockwell International)*

mechanical devices. Additionally, gas turbine meters adapt readily to electromechanical, fully electronic measurement systems.

As we have seen, both orifice meters and turbine meters are velocity-sensing devices and the accuracy of the derived volumetric measurement is directly dependent on the accurate sensing of the true velocity of the flowing gas stream through a known cross-sectional area.

Consider the following formula:

$$Q = V \times A$$
where Q = flow rate in cfh (cubic feet per hour)
$$V = \text{velocity}$$
$$A = \text{area}$$

The driving force for a gas turbine meter is the kinetic energy of the flowing gas. Kinetic energy may be defined as the physical energy of mass in motion.

The formula for determining kinetic energy is shown below:

$$KE = 1/w \, MV^2$$
where KE = driving force
$$M = \text{mass}$$
$$V = \text{velocity (flow rate)}$$

Note that when the KE factor is maintained a constant, any increase in one factor (M or V) permits a decrease in the other factor (V or M).

Theoretically, a turbine rotor mounted in a friction-free atmosphere would rotate when a single gas molecule impinged on a rotor blade regardless of the velocity of the gas molecule (see figure 7–15A).

As a practical matter, mechanical friction in the supporting bearings and gears of a gas turbine meter requires a minimal amount of kinetic energy to overcome this mechanical friction and cause rotor rotation at a speed directly proportional to the gas flow rate (see figure 7–15B). In addition to this mechanical friction, there is a fluid friction caused by the gas flowing through the passages of the meter, which also adds to the base kinetic energy requirement of the meter.

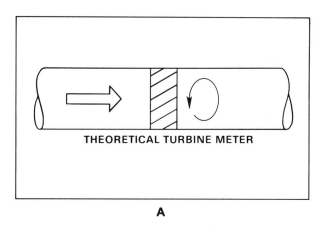

THEORETICAL TURBINE METER

A

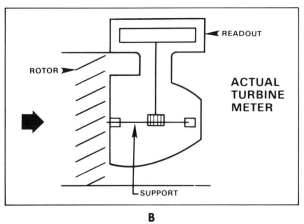

B

Figure 7–15 Turbine Meter Operation (*Courtesy, Rockwell International*)

Turbine meters have been used on practically every type of gas-measurement application ranging from the wellhead to the burner tip. On dry-gas wells, they are used for direct wellhead measurement.

Most commonly, production measurement initially involves a lot of condensates, water, and oil. In such cases, the turbine meter is installed downstream of a production separator and also on test separators.

Larger turbine meters are used in gathering systems and on transmission lines.

The widest application of gas turbine meters has been by the distribution utilities in metering large-volume industrial end-users.

A TURBINE FLOWMETER

The Electronic Flo-Meters, Inc. turbine flowmeter consists of a bladed rotor suspended in a flow stream with its axis of rotation perpendicular to the flow direction (see figures 7–16 and 7–17). It is a velocity-measuring device calibrated to indicate volumetric flow of liquid or gas in a pipe. The freely supported rotor revolves at a rate that is directly proportionally to the flow of the medium.

A magnetic pickup is used to sense the speed of the rotor. The pickup consists of a permanent magnet and a pickup coil. The pickup is mounted so that the rotor blades will cut the magnetic field. The rotor blades are made from a magnetic material so that as each blade cuts the field, a pulse is induced in the pickup coil. The output signal is a continuous sine-wave pulse train, with each pulse representing a discrete volume of the flowing medium. These pulses are

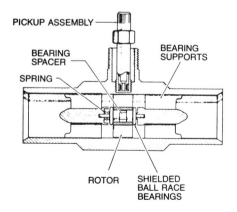

Figure 7–16   Small-Diameter Turbine Flowmeters *(Courtesy, Electronic Flo-Meters, Inc.)*

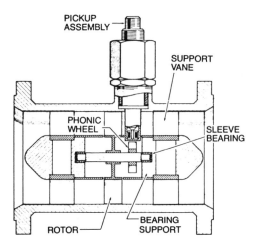

Figure 7–17   Large-Diameter Turbine Flowmeters *(Courtesy, Electronic Flo-Meters, Inc.)*

fed to appropriate electronic units for display of total flow volume or flow rate or to perform various control functions.

The flowmeters are constructed throughout using stainless steel. The rotor shaft is supported at both ends by sleeve bearings for liquid service and with prelubricated, shielded ball bearings for gas service. On meters 3 inches and smaller in size, the magnetic pickup will read the rotor blades directly. On meters 4 inches and larger, the pickup will read a phonic wheel, which is mounted on the rotor shaft.

AN INSERTION-TYPE FLOWMETER

Typical of a family of insertion-type flowmeters is the Electronic Flo-Meters, Inc. Model VR-600 illustrated in figure 7–18.

This unit is a completely automated insertion-type turbine flowmeter which incorporates the dual considerations of safety and precision in the measurement of high-pressure gas or liquid flows.

Insertion and retraction are accomplished hydraulically, using a reservoir pressurized either from the line or from an external source. The insertion/retraction operation may be performed locally with manually operated valving, or remotely by a signal from a control room or from pig switches located upstream and downstream of the meter. The turbine returns automatically to the proper depth and orientation.

Figure 7-18 An Insertion-Type Flowmeter *(Courtesy, Electronic FloMeters, Inc.)*

Characteristics of the insertion turbine meter include wide rangeability, negligible pressure drop, and the capability of servicing the equipment without line shutdown.

The unit is installed onto any line size, 4 inches and above, by a procedure involving the use of an isolation valve and a conventional hot tap device.

AN ULTRASONIC FLOWMETER

State of the art in ultrasonic flowmeters is the Sparling Model 500 (see figure 7-19). The Model 500 is designed for use in water and wastewater applications.

In its system, the unit is comprised of two interrelated components: (1) the sensor, which consists of a precalibrated, fused epoxy-coated flow tube containing a pair of externally mounted electro-acoustic transducers, and (2) a

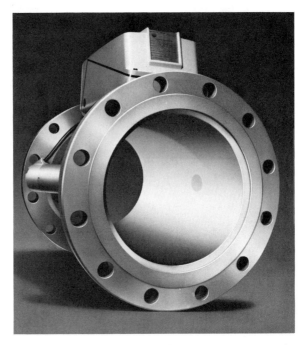

Figure 7-19   An Ultrasonic Flowmeter *(Courtesy, Envirotech-Sparling)*

transmitter or electronics package. The size of the flow tubes may be 4 inches to 48 inches, depending on the application.

The transmitter is packaged in a rugged NEMA 4 enclosure and is mounted either on the sensor, the pipe stand, or the wall. The transmitter and sensor can withstand accidental submergence in 30 feet of water for 48 hours and is designed for outdoor use.

Two outputs are provided. The prescaled pulse-rate output allows direct totalization without the expense and loss of accuracy of integrators (analog to pulse-rate converters). This pulse rate can also interface directly with electronic indicators, recorders, controllers and telemetering equipment, as well as pulse-input computer hardware.

The 4–20mA or 0–20mA output allows interface with electronic instrumentation that requires analog inputs.

In the event a transmitter needs replacement, the flow process does not shut down. Maintenance personnel simply replace the transmitter, which fits into an external well that is sealed off from the process fluid. Since the sensors are flush with the walls, there is no insertion loss. Furthermore, there is no blockage in the lines.

The transducers are mounted in an epoxy well so they will be isolated from the process system (see figure 7-20). This eliminates problems of corrosion and contamination of the transducer. Each system has a sensor with a fused epoxy-coated steel tube and flanges. A second transducer, serving as a transmitter, is made of epoxy painted and anodized cast aluminium with stainless-steel hardware.

The system utilizes a pair of ultrasonic transducers mounted to the exterior of the flow tube (see figure 7–21). Each transducer is capable of both

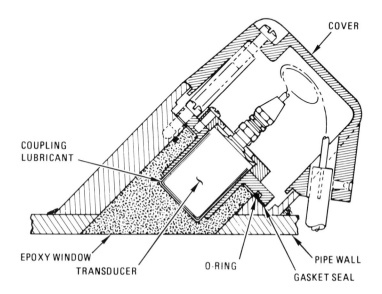

Figure 7–20 Cross-Section of Ultrasonic Flowmeter Transducer Mounting (*Courtesy, Envirotech-Sparling*)

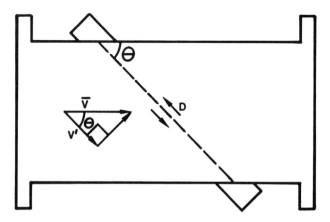

Figure 7–21 Operation of the Ultrasonic Flowmeter (*Courtesy, Envirotech-Sparling*)

sending and receiving ultrasonic pressure pulses. The flowmeter transmits pulses in two directions, against and with the flow, in order to make upstream and downstream time interval measurements.

The pulse transit times downstream (T_1) and upstream (T_2) can be expressed as follows:

$$T_1 = \frac{D}{C + V'} \text{ and } T_2 = \frac{D}{C - V'}$$

D is the distance between the transducers, C is the velocity of sound propagation in the fluid, and V′ is the fluid-velocity vector along the path of the pulses.

These two measurements are converted to two frequencies inversely proportional to time delay. The difference between the two frequencies is proportional to flow (average fluid velocity).

The units are insensitive to changes in the rate of propagation of sound. The output accuracy is unaffected by changes in fluid density, temperature, viscosity, or conductivity.

Signals are fed from the transducers to an electronic switch in the electronics package. The switch allows sharing of each transducer so that it can both send and receive signals. The signals are then directed to other electronics which amplify and condition the signal to derive a useable output. Output is digital or analog, whichever is the need.

MASS FLOW CONTROL

Mass flow control is a means of measuring and automatically controlling the weight-flow rate of a gas.

Controlling mass flow is the same as controlling the flow of molecules, since equal masses of a gas contain an equal number of molecules.

In contrast, volumetric flow measurement (including rotameters) and control must be corrected for local temperature and pressure conditions in order to determine molecular flow.

A Mass Flow Controller (See figure 7-22)

The figure is a photograph of the Tylan Corporation's Model FC-260 mass flow controller. The unit consists of a sensor assembly, a solid-state valve assembly, a printed circuitboard, a filtered inlet, a bypass assembly, and the housing.

The heart of the flow controller is a two-element sensor designed to

Figure 7–22 A Mass Flow Controller *(Courtesy, Tylan Corp.)*

dissipate minimum heat. It is encapsulated, sealed, and insulated to provide a near isothermal internal environment.

The solid-state valve is normally open, has no moving seals, no friction, and no mechanical wear. Operating on the principle of thermal expansion, the valve stem changes length by varying input power. As the length of the valve stem changes, its tip moves in and out of a seat, varying the gas flow. Unlike gear train, threaded stem, and modified solenoid-type valves, precise control is achieved because dead band and hysteresis are effectively eliminated.

The inlet has a fine-mesh 20-micron screen permanently installed in the inlet fitting to protect against contamination.

The bypass assembly precisely splits the gas flow, directing a sample of gas through the sensor.

An upper and lower housing contains the controller parts. The upper housing contains the sensor, the printed circuit, and the solid-state valve. The base (lower housing) includes the bypass assembly, the inlet and outlet fittings, and the valve seat.

The temperature rise of a gas is a function of the amount of heat added, the mass flow rate, and properties of the gas being used (see figure 7–23). To measure gas-flow rate, each flow instrument uses a small, stainless-steel sensor tube. Two external heated resistance thermometers are wound around the sensor tube.

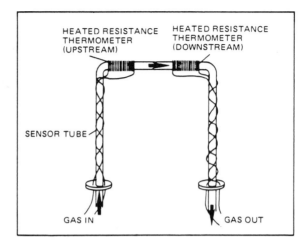

Figure 7-23   Operation of the Mass Flow Controller *(Courtesy, Tylan Corp.)*

When gas is flowing, heat is transferred along the line of flow to the downstream thermometer, thereby producing a signal proportional to the gas flow. The higher the flow, the greater the differential between the thermometers.

Each resistance thermometer has a power dissipation of 40 milliwatts and forms a part of a bridge and amplifier circuit that produces a 0 to 5Vdc signal proportional to the gas flow.

This signal is compared to a command voltage from a potentiometer (or other voltage source). From this comparison, an error signal is generated which adjusts the valve power to change the gas flow until the commanded setpoint is reached.

Each controller uses a thermal expansion design, normally open-valve. The valve is activated by a small, thin-walled tube with a ball welded on one end. This tube contains a heat-transfer fluid and small resistance heater element. When voltage is applied to this heater element, it causes the tube to expand relative to the outer shell, thereby moving the ball and controlling the flow. The solid-state valve has no moving seals, no friction, and essentially no moving parts.

A LIQUID LEVEL SYSTEM (See figure 7-24)

The Computer Instruments Corporation Model 7600 Bubbler System is typical of a liquid level system.

Bubbling a small flow of gas (usually air or any other compatible gas) from a submerged dip tube causes a gas back-pressure that is equal to the liquid

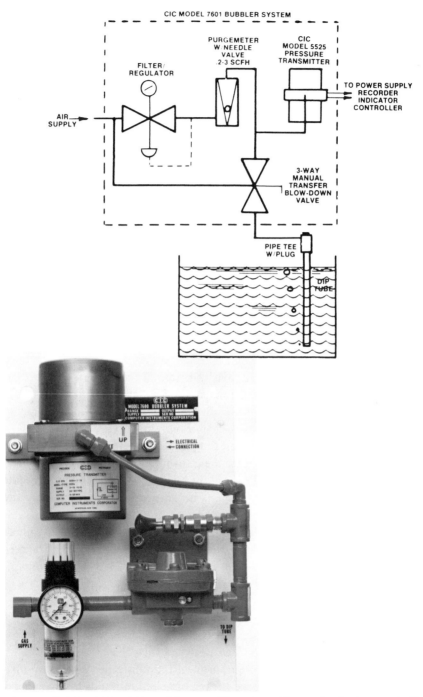

Figure 7-24   A Liquid Level System *(Courtesy, Computer Instruments Corp.)*

hydrostatic pressure; the gas pressure measured by the pressure transmitter indicates the liquid level.

The unit is especially suited for installations involving temperature, dangerous fluids, and slurries. Only the dip-tube material is critical when liquid is corrosive. Clearance for sediment should be provided between dip tube and tank bottom.

Most often used in open tanks, purge systems can be used in closed (pressurized) tanks if a differential pressure transmitter is used. The low-pressure port senses the tank vapor pressure, and relief-valve provision is made for the release of accumulated purge air. This system should not be used in elevated tanks where siphon action could back liquid into air lines if air supply were lost.

Typical bubbler systems in an open tank use gage transmitters. In a closed tank, the system uses a differential transmitter. The blowdown valve, when in the purge position, connects the air supply directly to the dip tube, thereby bypassing the transmitter. In the READ position, bubble air and the transmitter are connected to the dip tube.

Bubbler-System Applications   Several applications of the bubbler system are illustrated in figure 7–25. Operation is thus: The back-pressure in a small stream of air or gas escaping through the submerged open end of a dip tube is equal to the liquid head above it. A transmitter converts this back-pressure into an electrical signal proportional to liquid level and / or flow. The bubbler system is used to measure almost any fluid, including those which are highly corrosive, viscous, hot (molten metal), explosive, slurry type, or foodstuff.

In the first application the bubbler system is used to determine the tank level of a liquid. In this case the transmitter output is proportional to the level.

In the center application, the bubbler system is used for monitoring sewer flow. Here, flow in the pipe is obtained through ascertaining the pipe size, slope, material, and flow depth.

In the lower application, flow in an open channel is obtained when flow depth is known and using the standard formula for a submerged weir in a channel.

PRESSURE TRANSDUCERS

A pressure transducer is a device that responds to liquid or gas pressure and converts the pressure into an electrical variable.

Within the pressure transducer is a unit called a *pressure force summing device*. This device detects the pressure and responds by changing the pressure to a physical displacement. The physical displacement, in turn, causes an electrical transduction.

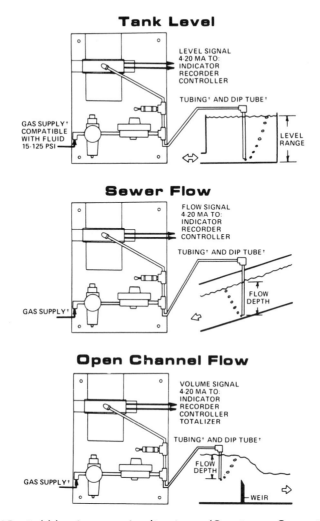

Figure 7–25 Bubbler-System Applications *(Courtesy, Computer Instruments Corp.)*

In figure 7–26, several of these summing devices are illustrated. The diaphragm device reacts to pressure applied to a simple diaphragm. The convoluted (also called corrugated) diaphragm and the capsule are variations of the simple flat diaphragm, but require a change in pressure from the source or measurand, as it is called.

The entities involved in pressure-force summing are the *mass, spring constant,* and the *natural frequency.* The frequency response of a pressure device is an extremely important parameter used in the selection of a pressure transducer. Below, the formula represents the natural frequency in a system in

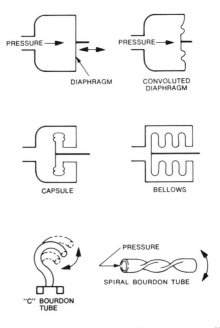

Figure 7–26   Pressure-Sensing Devices *(Courtesy, Bell & Howell CEC Division)*

which there is but one degree of freedom and with a constant diaphragm thickness:

$$f_n = \frac{1}{2\pi} \sqrt{\frac{K}{M}}$$

where f_n = natural frequency

K = spring constant

M = mass

It must also be realized that the electrical elements may contribute to this base formula along with sensitivity to vibration or acceleration. More information on frequency response will be provided later in the chapter.

The bellows (a convoluted diaphragm) illustrated in figure 7–26 responds to pressure by collapsing or expanding. Since the output arm is tied to the bellows, the output is representative of the input pressure.

The bourdon tubes illustrated in the drawing respond to pressure in a circular motion and a twisting motion, respectively.

Of course, the sensing element (summing device) is important in the performance of the transducer. However, we must not forget that good performance includes the electrical transduction element and the case installation. Force summing can be converted into a variety of different electrical devices. In general, the type of transduction sensor and its pressure summing device are determined by the amount of mechanical travel involved. The relationship of the sensor and the summing device must be ultimately compatible. That is, their marriage must not be made by personal choice of the designer but by the cooperative efficiency within the overall system.

Most pressure transducers are based on the following methods or concepts of transduction:

1. Capacitance
2. Inductive
3. Piezoelectric
4. Potentiometric
5. Strain gage

These concepts will be covered in detail further on in the chapter.

TRANSDUCER PRESSURE-RANGE DESIGNATIONS

Absolute Pressure   An absolute pressure transducer is referenced to a vacuum. One side of the transducer diaphragm is evacuated (reference vacuum). The other side is acted upon by the physical force (measurand). Absolute pressure is measured in pounds per square in absolute (psia).

Gage Pressure   A gage pressure transducer is referenced to atmosphere. One side of the transducer diaphragm is vented to atmosphere. The other side is acted upon by the physical force (measurand). Gage pressure is measured in pounds per square inch gage (psig).

Sealed Gage Pressure   A sealed gage pressure transducer is referenced and sealed to atmospheric pressure with a partial gas mixture such as helium. One side of the diaphragm is sealed. The other side is acted upon by the physical force (measurand). It is used at ranges where atmospheric pressure changes do not affect measurements. It is also used to prevent the ambient medium from entering the transducer case. Sealed gage pressure is measured in pounds per square inch gage (psig) as referenced to sea level.

Differential Pressure   Differential pressure transducers have two pressure ports incorporated to permit application of pressure to both sides of a diaphragm. This permits measurements of small pressure differences of high line pressures with accuracies much superior to those transducers which monitor pressure independently. Two separate sensors may be placed in one case and their difference monitored electrically. Differential pressure is measured in pounds per square inch differential (psid).

Static Overpressure   Static overpressure can be designated in three terms: overpressure that does not cause a calibration shift, overpressure that does not cause permanent damage, and overpressure that does not burst the transducer case.

Dynamic Overpressure   Dynamic pressures are transient overpressures beyond the specified range that may cause damage or calibration shifts. These overpressures may be different than the same level of pressure for longer durations. Short-interval overpressure transients may exist without being apparent to the user.

Linearity   Linearity in a pressure transducer has to do with the closeness of the transducer's calibration curve to a predetermined straight line. With these curves, the transducer output can be predicted and corrected. Computers make this a simple operation.

Hysteresis   Hysteresis is the maximum difference in output at any input (measurand) within a range of pressures, when the pressure value is approached first with increasing and then decreasing measurand. Maximum hysteresis is the difference between mid-scale reading taken on increasing to full scale and decreasing from full scale toward zero.

Repeatability   This is the transducer's ability to reproduce identical outputs when the same measurand is applied under the same variables or conditions.

Stability   Stability is the transducer's ability to continue its performance characteristics over time and temperature.

Drift   Drift is a shift from normal transducer output over time.

Frequency Response   Frequency response is the transducer's predicted output response to a given input pressure. Frequency response is also a ratio of the output and the measurand.

Frequently, pressure measurement is used to establish time-event sequences. Response to pressure changes can be affected by the type of transducer and the volume of its pressure chamber, as well as by interconnecting plumbing.

Volume displacement of the pressure-sensing cavity with change in measurand pressure can be significant when small or long tubing is employed. Tube length and diameter affect the transducer's response to step changes.

PRESSURE TRANSDUCER SELECTION

The selection of a pressure transducer evolves around the pressure (measurand) environment. The selector must be aware of what effects will occur before, during, and after measurement. There are essentially three direct pressure applications. These are illustrated in figures 7–27 through 7–29. The first, in figure 7–27, is an *absolute pressure transducer*. This transducer is referenced to a vacuum. Typically, the interior of the transducer case is evacuated and sealed. Some absolute pressure transducers are referenced to a bellows or a capsule

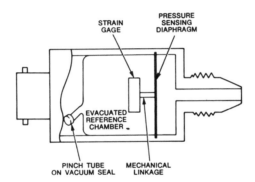

Figure 7–27   Absolute Pressure Transducer *(Courtesy, Bell & Howell CEC Division)*

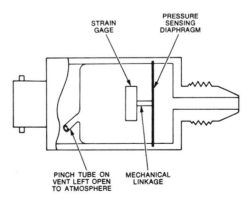

Figure 7–28   Gage Pressure Transducer *(Courtesy, Bell & Howell CEC Division)*

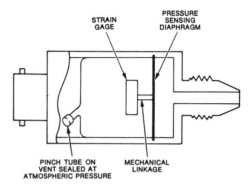

Figure 7-29 Sealed Gage Pressure Transducer *(Courtesy, Bell & Howell CEC Division)*

whose exterior is acted upon by the measurand (pressure). In figure 7-28, the case is vented to atmosphere rather than being evacuated. Other than the vent, the *gage pressure application* is the same as absolute pressure transducer.

In figure 7-29, *sealed-gage pressure application* is presented. This transducer is designed to prevent ambient media from entering the transducer case. Usually these instruments are of a range in which atmospheric pressure does not affect the measurement. The application is generally built with a partial atmosphere of helium sealed within. The application is called a sealed gage pressure transducer.

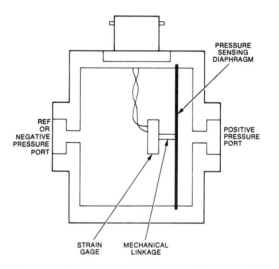

Figure 7-30 Differential Pressure Transducer *(Courtesy, Bell & Howell CEC Division)*

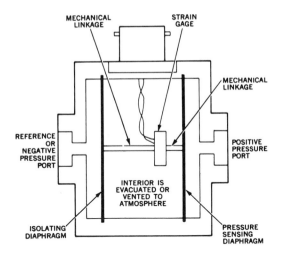

Figure 7–31 Isolated, Evacuated, or Vented Differential Pressure Transducer *(Courtesy, Bell & Howell CEC Division)*

There are three differential-pressure transducer applications. In figure 7–30, the basic *differential pressure transducer* is represented. The device has two parts. This allows application of pressure to both sides of the pressure-sensing element. The pressure-sensing diaphragm senses the pressure difference

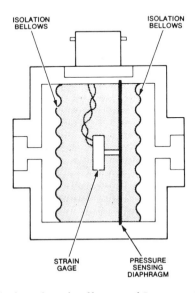

Figure 7–32 Oil-Filled, Isolated Differential Pressure Transducer *(Courtesy Bell & Howell CEC Division)*

between the positive sensing part and the negative or reference part. The diaphragm is connected mechanically to a strain gage and, in turn, electrically to an output connector.

Figure 7–31 is an *isolated differential pressure transducer.* There are two diaphragms included within the case of this transducer. The interior is evacuated or vented to atmosphere. Pressure from the positive sensing part is applied to the pressure-sensing diaphragm. Pressure from the reference or negative part is applied to the isolating diaphragm. A mechanical linkage exists between the two diaphragms. As in the other applications, the strain gage is electrically tied to an output connector.

In some cases, the designer fills the interior of the case with oil (see figure 7–32). This allows the transmission of high line pressures throughout the instrument, up to the pressure-sensing element located at the pressure side of the transducer case. The *oil-filled differential transducer* is expensive to build as compared to the wet/dry configuration.

THE CAPACITIVE PRESSURE TRANSDUCER

The capacitive transducer consists of two fixed conductive plates which are isolated from a housing by insulated standoffs (see figure 7–33). A pressure port directs pressure into a bellows. Attached to the bellows is a diaphragm. As pressure changes, the bellows expands or retracts, changing the position of the diaphragm, which serves as one capacitor plate. This causes a capacitance change in two separate capacitive circuits. In some transducer designs, only a single capacitive element is employed. The change in capacitance is used to vary the frequency of oscillators or to null a capacitance bridge. The dual concept

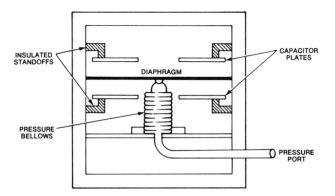

Figure 7–33 Capacitance Pressure Sensor *(Courtesy, Bell & Howell CEC Division)*

allows two oscillators to operate in a highly linear mode. Small displacement of the capacitor diaphragm is a major advantage. Capacitance sensors may be used for pressure transducers in ranges from 100 psi and down to medium vacuum ranges.

The capacitive transducer has excellent frequency response, standard hysteresis, repeatability, and stability with excellent resolution. Its disadvantages include a high impedance output with additional electronics. Capacitor leads from the sensor must be short to eliminate stray pickup. Finally, the capacitive transducer is extremely sensitive to temperature variations.

A Typical Capacitive Pressure Transducer (See figure 7-34) The Setra Systems, Inc. Model 239E is typical of a family of capacitive pressure sensors.

The unit is designed for use with a digital pressure readout / indicator. The pressure media may be compatible liquid or gas. The reference pressure must be clean, dry air or noncondensable gas. The high output signal, excellent stability, fast dynamic response, and lower price make this transducer well suited for many industrial, laboratory instrumentation, and aerospace applications.

The construction of the capacitive pressure transducer is illustrated in figure 7-35. A description of its operation follows.

In the pressure transducer shown in simplified form in figure 7-35A, a low-pressure sensor is illustrated. Low-pressure-range sensors have a thin

Figure 7-34 A Capacitive Pressure Transducer *(Courtesy, Setra Systems Inc.)*

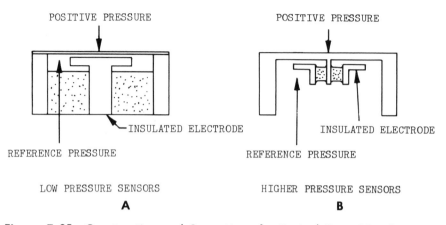

POSITIVE PRESSURE   POSITIVE PRESSURE

INSULATED ELECTRODE   INSULATED ELECTRODE

REFERENCE PRESSURE   REFERENCE PRESSURE

LOW PRESSURE SENSORS   HIGHER PRESSURE SENSORS
        A   **B**

Figure 7–35 Construction and Operation of a Typical Capacitive Pressure Transducer *(Courtesy, Setra Systems Inc.)*

stretched stainless-steel diaphragm positioned close to the electrode. Positive pressure moves the diaphragm toward the electrode, increasing the capacitance. High positive overpressure pushes the diaphragm against the electrode, thereby providing high positive overpressure protection.

In figure 7–35B, a higher-pressure sensor is shown. Higher-pressure-range sensors have an insulated electrode fastened to the center of the diaphragm, forming a variable capacitance. As the pressure increases, the capacitance will decrease.

The capacitance is detected and converted to a linear dc electrical signal.

THE DIFFERENTIAL TRANSFORMER

The differential transformer consists of a primary and secondary transformer with a magnetic movable core. One end of the core is attached to a push rod. The push rod is mechanized by pressure-sensing diaphragm, bellows, or bourdon tube (see figure 7–36). As pressure changes, the push rod displaces the magnetic core within the transformer. As the core is displaced, an imbalance is produced within two secondary windings. AC excitation is usually 50 to 60 hertz, but 10,000 hertz is employed to reduce the size and mass of sensor components. The sensitivity of the transducer is a tradeoff in design between the core and the turns ratio. Owing to the relatively large core, the transducer cannot be used in areas of high vibration or acceleration.

A Typical LVDT Pressure Sensor (See figure 7–37) The Computer Instruments Corporation Model 5000 is typical of the many LVDT pressure

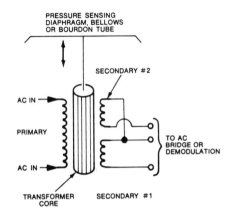

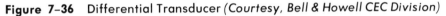

Figure 7–36 Differential Transducer *(Courtesy, Bell & Howell CEC Division)*

transmitters on the market. It is a general-purpose device designed for gage, absolute, and differential pressure applications where only limited overpressure ratings are required.

The unit utilizes an LVDT (Linear Variable Differential Transformer) as part of a simple, rugged pressure-sensing mechanism. The mechanism is incorporated into a transmitter package designed for convenience and flexibility in use and capable of performing under severe environments.

In all cases, the pressure media encounters only metal parts and is completely isolated from the electronics and electrical connections; corrosive or dirty environments are tolerated. Suitable isolating diaphragms and filters

Figure 7–37 An LVDT Pressure Transmitter *(Courtesy, Computer Instruments Corp.)*

should be employed. Drain plugs are provided to permit through-flushing of all pressure chambers.

The LVDT pressure-sensing system (see figure 7–38) does not utilize any levers, gears, or other linkages. It consists simply of a magnetic core fastened directly by a short stem to a metal capsule. Pressure entering through the "HI" pressure port causes expansion of the capsule and corresponding movement of the core in the LVDT, producing a known change in the LVDT electrical output. Depending upon the particular electronic circuitry supplied internally, the unit provides voltage or current outputs precisely proportional to the pressure input.

Pressure applied through the "LO" pressure port enters a metal isolating chamber surrounding the capsule and acts to compress the capsule (opposes "HI" pressure core motion). Thus, when pressures enter both "HI" and "LO" ports simultaneously, the unit's output becomes precisely proportional to the pressure difference. In the case of low-range differentials, proper operation requires careful filling procedures to eliminate trapped air.

As a "gage" unit, the isolating chamber is omitted and the case is vented to the atmosphere through the "LO" pressure port; input pressure is connected only to the "HI" pressure port. When used as an "ABSOLUTE" unit, the cap-

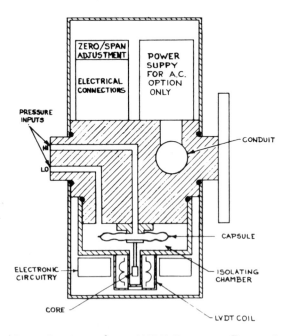

Figure 7–38 Cross-Section of an LVDT Pressure Transmitter *(Courtesy, Computer Instruments Corp.)*

sule is evacuated and sealed and input pressure is applied only to the "LO" pressure port.

POTENTIOMETRIC PRESSURE TRANSDUCER (See figure 7–39)

A basic potentiometric pressure transducer has a force summing pressure bellows attached to the linkage. The linkage is mechanically connected to a potentiometer wiper. The wiper travels over a multiturn wire coil or deposited resistor, as the sensor reacts to pressure changes.

Usually some motion amplification between the force summing element (pressure sensor) and the wiper track across the resistance element is employed to minimize acceleration error.

The resistance of the potentiometer can provide specialized output linearities such as linear, sine, cosine, exponential, etc. As a voltage ratio device, close regulation of the excitation voltage is not required. High-level output is inherent in the potentiometer concept. Output load impedance must be kept high to limit loading effects.

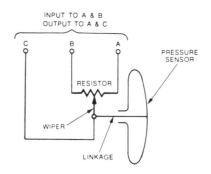

Figure 7–39 Function of a Potentiometric Sensor (*Courtesy, Bell & Howell CEC Division*)

A Typical Potentiometric Pressure Transducer (See figure 7–40)   The Vernitech Model 9000 is typical of the potentiometric pressure transducers. This transducer utilizes a bourdon tube sensing unit. Pressure ranges are 0 to 100, 0 to 5000, 0 to 1000, and 0 to 5000 psig. Potentiometric resistor values are in the range of 1K, 2K and 5K ohms. The unit has a pressure port on one end and potentiometer electrical terminals on its face. The case is made from aluminum phenolic. The pressure fitting is stainless steel. Operation of the unit is as follows: in the low-pressure ranges (figure 7–41A), the sensing element is a metal capsule consisting of a pair of symmetrical concentrically corrugated diaphragms. The diaphragms are welded together at their outer rims to form a

Figure 7–40 A Typical Potentiometric Pressure Transducer *(Courtesy, Vernitech)*

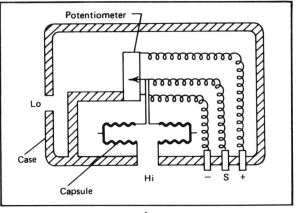

A

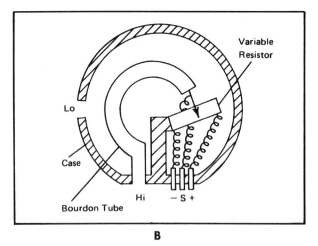

B

Figure 7–41 Low- and High-Pressure-Range Potentiometric Transducers *(Courtesy, Vernitech)*

hollow flexible member. The member has a predictable motion when fluid or gas pressure is applied internally or externally. In the high-pressure ranges (figure 7–41B), the sensing element is a metal Bourdon tube. The tube consists of a circular tube of oval cross-section, brazed closed at one end. The pressure applied internally or externally causes the circular tube to partially straighten and its closed tip to move in a predictable path. In all cases the electromechanical transducer is a variable resistor, in which the movable contact or slider is driven by the motion of the capsule or Bourdon tube. When voltage is applied to the ends of the variable resistor, the movable contact position appears on the output terminal as a proportionate fraction of the excitation voltage.

An absolute pressure transducer provides an output voltage that is proportional to the difference between the applied pressure and a perfect vacuum. In such a unit, the pressure-sensing element is completely evacuated and sealed; the HI pressure port is not present and input pressure is through the LO port.

A gage pressure transducer provides an output voltage that is proportional to the difference between the applied pressure and the ambient pressure. In such a unit, the input pressure is through the HI port and the ambient pressure is applied through the open LO port.

A differential pressure transducer provides an output voltage that is proportional to the difference between two applied pressures. In such a unit, the higher of the two pressures is applied through the HI port and the lower through the LO port.

THE PIEZOELECTRIC PRESSURE TRANSDUCER

The Piezoelectric Effect   The piezoelectric effect is the property of a crystal exhibited by the generation of a voltage when pressure is applied. A force exists between atoms inside a given material because of the interlaced electronic orbit and mutual repulsion between charged nuclei. Many elements and compounds have atoms that are assembled in a pattern known as a *crystalline structure.* The structure provides stability and equilibrium to the material. Individual atoms vibrate and their electrons follow interlacing orbits but stay in their same relative position. In most crystalline structures, the formation resists change caused by heat, electricity, pressures, etc. Some substances such as tourmaline, Rochelle salt, and quartz are pressure-sensitive and release electrons when the crystal cubes are compressed along specific lines or axes. The subject materials are inherently piezoelectric. That is, under pressure, they will vibrate at a very specific frequency and have a stable output for long periods of time. Other materials are not naturally piezoelectric but can be transformed into a quasi-piezoelectric crystal by poling. Poling is the momentary application of a strong direct current. Such is the case of barium titanate and lead zirconate–lead titanate. These are ceramic materials.

The Quartz Resonator (See figure 7–42)   A quartz resonator using the piezoelectric effect was developed by the Kearfott Division of the Singer Company as a digital force sensor in the mid-1960s. Since that time, this quartz sensor has been widely used as the basic pickoff device for pressure transducers manufactured by Paroscientific, Inc.

The first part of figure 7–42 is a line drawing of a quartz resonator. The reader will first note that the entire resonator is fabricated from a single piece of quartz. This eliminates losses caused by joints and connections. The resonator is fastened to a structure that can transmit force to it. An integral mounting

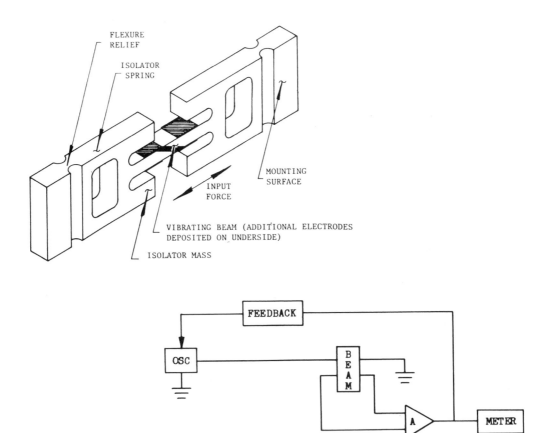

Figure 7-42   The Quartz Resonator *(Courtesy, Paroscientific Inc.)*

isolation system decouples the vibrating beam from the force-producing structure. These isolation springs and masses vibrate at a much lower frequency than the vibrating beam, thereby acting as a low-pass mechanical filter. Very little energy is transmitted from the beam to the mounting surfaces, resulting in high "Q" operation.

The resonant frequency of the sensor is determined by its dimensions, composition, and stress load. Under tension, the frequency increases. Under compression, the frequency decreases. The resonant frequency is thus a true and accurate measurement of the applied load.

In the second part of figure 7–42, a total system is illustrated. The beam, under tension, increases frequency as referenced to an oscillator. The difference is amplified and displayed. A feedback line is provided to the oscillator.

An Absolute Pressure Transducer Using the Quartz Crystal as a Sensor (See figure 7–43)   A typical absolute pressure transducer design using the quartz resonator is produced by Paroscientific, Inc.

The figure is a cross-section of the design. The transducer consists of the quartz crystal, a single bellows, a pivot, and balance weights.

In operation, an absolute pressure is applied to pressure input P_1. A high internal vacuum is used as a reference within the housing. The bellows exerts a force proportional to the product of the absolute pressure times the bellows' effective area. When the bellows and the quartz beam are equidistant from the pivot, the same force is applied to the beam as is produced by the bellows. By choice of bellows' effective area and lever-arm distances, different pressure

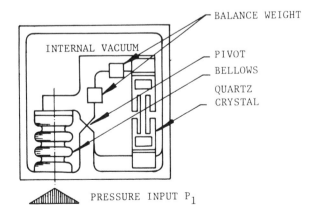

Figure 7–43   An Absolute Pressure Transducer Using a Quartz Crystal as a Sensor *(Courtesy, Paroscientific Inc.)*

ranges can be obtained within the same frequency range and stress levels for a given quartz resonator.

The pivotal arrangement allows a counterbalance system in which balance weights are adjusted in size and position to ensure that the center of gravity is coincident with the pivot point. This adjustment makes the transducer insensitive to linear accelerations, since there is no moment arm for the forces to act upon.

A Differential Pressure Transducer Using the Quartz Crystal as a Sensor (See figure 7-44) A typical differential pressure transducer design using the quartz resonator is produced by Paroscientific, Inc.

The figure is a cross-section of the design. The transducer consists of the quartz crystal, a double bellows, a pivot, and balance weights.

In operation, pressure is applied to pressure inputs P_1 and P_2. The quartz resonator operates in a high internal vacuum within the housing. With the pressure applied through both inputs to their respective bellows, two coaxial and counteracting forces will be applied to the lever arm. If P_1 and P_2 are the same value, no net force results. If P_1 is different than P_2, only the difference force will be applied to the quartz beam. When the bellows and the quartz beam are equidistant from the pivot, the same force is applied to the beam as is produced by the bellows. By choice of bellows' effective area and lever-arm distances, pressure ranges can be obtained within the same frequency range and stress levels for a given quartz resonator.

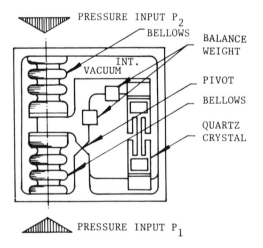

Figure 7-44 A Differential Pressure Transducer Using a Quartz Crystal as a Sensor *(Courtesy, Paroscientific Inc.)*

As in the absolute pressure transducer, balance weights are adjusted in size and position to ensure that the center of gravity is coincident with the pivot point. This adjustment makes the transducer insensitive to linear accelerations, since there is no moment arm for the forces to act upon.

Piezoelectric Excitation (See figure 7–45)   The most prevalent method of driving a vibrating beam at its resonant frequency is by way of piezoelectric excitation. In figure 7–45 the response of piezoelectric excitation is illustrated. A crystal of quartz is exposed to an electric field. In the figure a characterization of the excitation is shown. Four electrodes are vacuum-deposited on the vibrating beam in a manner to ensure that the electrode polarities are diagonally opposed. The vibrating beam is forced into flexural vibration by an oscillator circuit which tunes itself to the vibrating beam's resonant frequency.

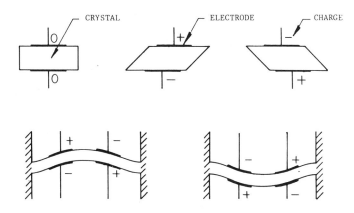

Figure 7–45   Piezoelectric Excitation *(Courtesy, Paroscientific Inc.)*

Typical Quartz Pressure Transducers (See figure 7–46)   Typical of the quartz pressure transducers are the Paroscientific Models 245-A and 230-D. These are illustrated in figure 7–46 along with a quartz crystal. Model 245-A is an absolute pressure transducer, while Model 230-D is a differential pressure transducer.

The sensors, in use with a shock mount, are capable of withstanding extremely high acceleration, shock, and vibrational loads.

The quartz pressure transducers may be applied in meteorology, oceanography, air data systems, jet engine testing, propulsion-control systems, wind-tunnel instrumentation, transfer pressure standards, and energy exploration.

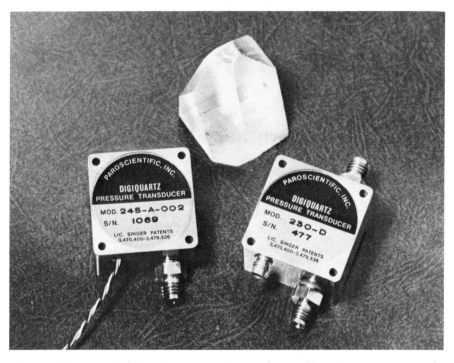

Figure 7-46   Typical Quartz Pressure Transducers *(Courtesy, Paroscientific Inc.)*

THE STRAIN-GAGE PRESSURE TRANSDUCER

The strain-gage pressure transducer has a solid-state sensor as its primary element of construction. Pressure is converted to strain using pressure-sensitive solid-state material. The strain-sensitive materials provide an electrical output that is proportional to the pressure.

Figure 7-47 is a strain-gage pressure transducer. The sensor is located between the reference chamber and the temperature compensator. The basic sensing element consists of a four active arm bridge of thin-film strain gages applied by sputtered-film deposition. For pressure ranges of 1,000 and above, the gage elements are applied directly to the pressure diaphragm. For the lower pressure ranges, the gage elements are applied to a dual cantilever beam welded to the pressure diaphragm.

The basic sputtered sensor is electron-beam-welded to the pressure chamber/adapter, which also provides a high degree of mechanical isolation from mounting torque effects. The sputtered sensor is also well isolated from external case effects, since the case is welded to the pressure chamber/adapter

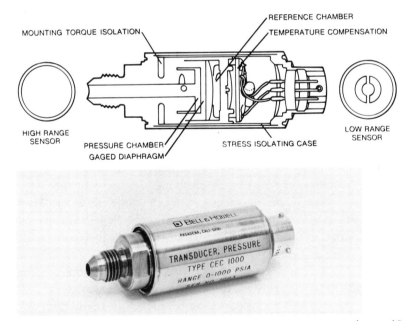

MOUNTING TORQUE ISOLATION

REFERENCE CHAMBER

TEMPERATURE COMPENSATION

HIGH RANGE SENSOR

LOW RANGE SENSOR

PRESSURE CHAMBER
GAGED DIAPHRAGM

STRESS ISOLATING CASE

Figure 7–47 Cross-Section of a Strain-Gage Pressure Transducer *(Courtesy, Bell & Howell CEC Division)*

and is not in contact with any portion of the sensor. The next section of this chapter covers the history and process of sputtering.

The primary advantage of using a sputtered-gage pressure transducer is its stability with time over a broad temperature range.

History of Sputtering

Sputtering is not a new concept; the phenomenon was first described in 1852 by Sir William Robert Grove, who referred to the process as "cathode disintegration."

The first reported commercial application of sputtering did not take place until 1928. Western Electric Company used cathode disintegration for the manufacture of phonograph records and contacts for microphone transmitters.

The process has been refined considerably during the past decade, and its use has been extended to include deposition of dielectric materials. It is widely used in modern industrial production processes for the deposition of dielectric thin films used in microcircuitry. These thin-film techniques are rapidly replacing the less controllable vacuum deposition processes used previously.

Applying the sputtering process to the manufacture of a strain-gage-pressure transducer is new. It is the first technological advance in the strain-gage field in the past several years. The unique construction method offers many performance advantages.

The Sputtering Process

Sputter deposition takes place in a vacuum chamber.

During sputtering or cathode disintegration, molecules of the gage and insulating material are ejected from an electrode held at a negative potential by the impact of positive gas ions bombarding the surface.

The ejected molecules strike the target area with kinetic energy several orders of magnitude greater than any other deposition method. The high energy impact of the molecules creates the superior adherence associated with the sputtering process.

To obtain the necessary ionization, a gas discharge is maintained between the anode and cathode (target). An inert gas (such as argon) is continuously introduced into the vacuum chamber to provide the discharge. The pressure in the chamber is maintained in the range of 10 to 1,000 microns. The spacing between anode and cathode is typically 2 to 3 inches, with a potential between them in the range of 1,000 to 10,000 volts.

In the equilibrium state, an electron is emitted by the cathode and is accelerated toward the anode. The electron, by collision with the argon molecules, produces positive ions that strike the cathode and eject another electron. The gas ions are accelerated with enough energy to allow the bombardment of the cathode to actually physically displace molecules of material. These molecules are accelerated toward the anode and impinge upon it with the force of several thousand electron volts, as opposed to simply condensing on the surface.

The cathode is composed of the material determined to be ideal for the application. Cathode disintegration permits an unlimited choice of gage and substrate materials.

The basic sputtering process described is called *diode sputtering*. It can be used for sputtering conductive materials, but cannot be used for dielectrics due to the buildup of a surface charge on the cathode which stops the sputtering process. In order to sputter dielectric materials such as the insulating layer in a thin-film strain gage, rf sputtering is employed.

Manufacturing the Sputtered Strain Gage (See figure 7-48)   To produce a sputter-deposited strain-gage pressure transducer, the metallic surface of each

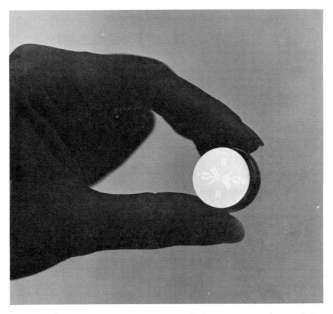

Figure 7-48 Gage Pattern on a Sputtered-Gage Transducer *(Courtesy, Bell & Howell CEC Division)*

strain diaphragm first must be highly polished. An extreme polish is required, since the dielectric substrate layer is deposited as a thin film less than one-half-thousandth of an inch thick. Any surface defect would penetrate the thin layer and cause a short of the strain-gage elements.

Following the mechanical polish, the pressure diaphragms are arranged in the sputtering chamber and the system is evacuated. The remaining steps of the manufacturing process take place in a vacuum environment, and thus avoid exposure of the sensor elements to contaminants.

The diaphragm surfaces are further cleaned by sputter-etching a small amount of metal from each active face using a reversed potential. This prepares the surface for the thin insulating layer to be applied by rf sputtering.

Next, the dielectric insulating layer is deposited as a thin film over the entire diaphragm surface. No mechanical masking is needed. After the dielectric insulation has been applied, the first cathode is moved aside and the gage cathode (material source) is positioned. Both sources are present in the chamber throughout the procedure, to accomplish deposition of both layers without exposing the sensors to ambient conditions.

Figure 7-48 is a gage pattern on a sputtered-gage transducer.

REFERENCES

Reference data and illustrations in Chapter 7 were supplied by state-of-the-art manufacturers.

Permission to reprint was given by the following companies:

Capacitive pressure transducers—Setra Systems, Natick, Massachusetts

Electronic flow meters—Electronic Flo-Meters (EFM) Inc., Dallas, Texas.

Gas meters—Rockwell International, Pittsburgh, Pennsylvania

Level and flow transducers—Computer Instrument Corp. (CIC), Hempstead, New York

Mass flow—Tylan Corp., Torrance, California

Piezoelectric pressure transducers—Paroscientific, Inc., Redmond, Washington

Potentiometric transducers—Vernitech Division, Vernitron Corp., Deer Park, New York

Pressure transducers, general—Bell & Howell, CEC Division, Pasadena, California

Strain-gage pressure transducers—Bell & Howell, CEC Division, Passadena, California

Ultrasonic flowmeters—Environtech Inc.–Sparling, El Monte, California

All copyrights © are reserved.

8

Electromagnetic Waves:
Ultrasonics and Microwave

Ultrasonic waves are vibrational waves of electromagnetic frequencies that are above the hearing range of the normal ear. The term includes waves of a frequency of more than 20,000 hertz. The presence of a medium is essential to the transmission of ultrasonic waves and almost any material that has elasticity can propagate ultrasonic waves. This propagation takes the form of a displacement of successive elements of the medium.

THEORY OF ULTRASONIC WAVE SYSTEMS (See figure 8–1)

A simple sound wave, as it travels outward from its source, loses strength rapidly as the distance increases (see figure 8–1A). This decrease in the strength of sound waves along a path can be greatly affected by discontinuities within the path (see figure 8–1B). With an ultrasonic control system, a sound path through air is established. The strength of the sound wave at any point along the path is a function of the distance from the point of origin. The introduction into the path of any material capable of absorbing some of the sound energy or reflecting it away from the original path can be measured. This change in the normal weakening or attenuation of sound along a path can be used to operate electronic circuitry.

A more thorough study of the properties of sound waves in air will demonstrate that other factors than distance cause attenuation of sound waves. Relative humidity effects, temperature effects, and the presence of standing waves are the major stumbling blocks to this type of a control system. Complicated electronic circuitry can be developed to circumvent most of these problems. However, this results in a device that is expensive to manufacture, critical to adjust, and difficult to maintain.

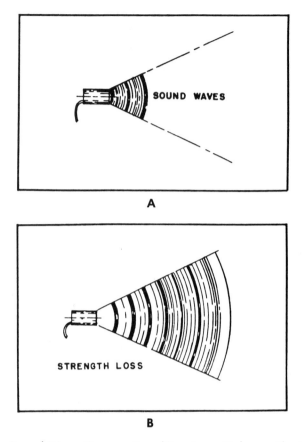

Figure 8-1   Sound-Wave Propagation *(Courtesy, Delavan Electronics, Inc.)*

THE TWO-SENSOR ULTRASONIC WAVE CONTROL SYSTEM

There are many ultrasonic wave control systems. Some are used in an air medium. Others are used in a media such as water or other liquids. Since it is impossible to cover all the types in a single chapter, we shall limit our discussion to the air medium.

The two-sensor ultrasonic system operates much like the acoustic howl or ringing sound heard when the microphone of a public address system is placed too close to the loudspeakers or when the volume control of the public address system is turned up too high (see figure 8–2A). Because at normal temperatures the gas molecules in the air are in motion, minute sounds are always present in the atmosphere. Additionally, electrical amplifiers produce some unavoidable electrical noise of their own.

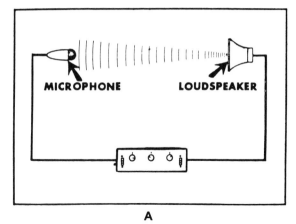

A

B

Figure 8-2 Simplified Ultrasonic Wave Control System *(Courtesy, Delavan Electronics, Inc.)*

When the extremely weak sounds issuing from the loudspeaker because of unavoidable electrical noise and random sounds found in the air reach the microphone, they are converted into electrical energy. They are then amplified by the amplifier and are issued from the loudspeaker with considerably greater strength. These strengthened sound waves traveling to the microphone are again reamplified and reissued at even greater strength. This process continues and in a fraction of a second builds up to the familiar howl or ringing. The particular tone or frequency of this ringing is determined by the characteristics of the amplifier, loudspeaker, and microphone, and to a large extent by the properties of the acoustic path in the air between the microphone and the loudspeaker.

In the Sonac system, one sensor is connected to the Sonac amplifier and operates essentially as a loudspeaker (see figure 8–2B). This transmitting sensor will produce ultrasonic sound waves of the quantity and frequency delivered to it by the amplifier. The other sensor, which we will call the receiving sensor, is connected to the amplifier as a microphone and will deliver to the amplifier as electrical energy and ultrasonic sounds reaching its diaphragm. Figure 8–3 is a photograph of the two-sensor ultrasonic system.

The ultrasonic sensors are directional in their response to sound waves. If the transmitting and receiving sensors are positioned facing each other, and the path between the two sensors is unobstructed, and the electrical gain in the amplifier is sufficient to overcome the losses in sound energy across the path between the two sensors, acoustic feedback will occur. This, of course, cannot be heard, because the sensor operates well above the range of human hearing. This condition of acoustic feedback can occur only when the electrical gain of the control amplifier is equal to or exceeds the loss in sound strength across the path between the sensors.

In a practical operating system, the electrical gain of the amplifier must be equal to or exceed the total path loss under any ambient environmental condition.

The object to be detected must be of sufficient size and be placed in the acoustic path in such a manner as to cause attenuation. Since some diffusion and refraction of sound occurs around the perimeter of the object to be

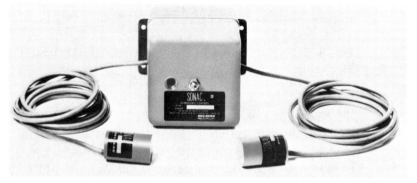

Figure 8–3 The Two-Sensor Ultrasonic Sensor System *(Courtesy, Delavan Electronics, Inc.)*

detected, the acoustic shadow formed by the object does not continue behind the object indefinitely. Locating the receiving sensor as close as mechanically possible to the object to be detected is desirable.

The Two-Sensor Ultrasonic System Installation (See figure 8–4)

The ultrasonic system can be used in four basic ways. First, the sensors can be installed to provide a direct path in which an obstruction between the sensors breaks the acoustic path, thereby operating a relay (see figure 8–4A). A relay may be connected to the desired external function (light, buzzer, switch, etc.). Although ultrasonic waves are transmitted from one sensor to the other sensor at approximately a 50° angle, an object large enough to cover the active area of the second sensor will break the beam for short path lengths.

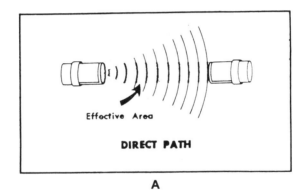

Effective Area

DIRECT PATH

A

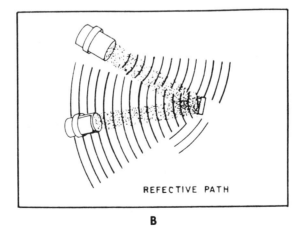

REFECTIVE PATH

B

Figure 8–4 Methods of Sensor Installation *(Courtesy, Delavan Electronics, Inc.)*

Second, the direct path method can also be used to sense the absence of an object from the beam. With the object obstructing the beam, the output relay is closed. When the object is removed, the acoustic path is completed, which opens the relay.

The third and fourth ways in which the sensor can be used are with sensors located to provide a reflective path (see figure 8–4B). As with the direct path, the reflective beam can also be used in two ways—with the relay open to sense objects or with the relay closed to sense the removal of objects from the beam.

In this method the sound waves emitted from one sensor are reflected from an object to the second sensor to provide the acoustic path. When the object is removed from the path, the relay is closed. Conversely, when objects establish a path, this opens the relay.

The sensor uses ultrasonic energy to produce a direct path or a reflective path.

The Two-Sensor Ultrasonic System Probe (See figure 8–5)

The probe is basically a magnetorestrictive device consisting of a diaphragm, nickel tube, magnet, and pickup coil.

When energy is supplied to the drive coil, it causes the diaphragm to vibrate. The frequency is determined by the mechanical resonant system within the probe. Electrical energy will then be transferred to the pickup coil if the

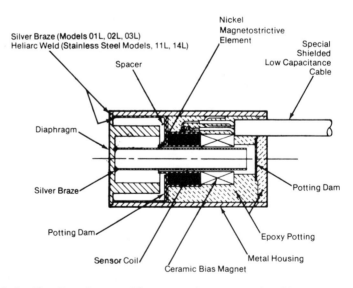

Figure 8–5   The Two-Sensor Ultrasonic System Probe *(Courtesy, Delavan Electronics, Inc.)*

diaphragm is free to move in gas. When the diaphragm motion is damped by a noncompressible liquid, no energy is transferred to the pickup coil.

Connecting the pickup coil of the probe to the input of an amplifier and the output of the amplifier to the drive coil of the probe forms a feedback-loop circuit in which any energy appearing in the output of the probe will be fed to the amplifier and there amplified and then furnished to the input of the probe. This causes vibrations to occur in the diaphragm and furnishes a signal back to the amplifier for reamplification. If the gain of the amplifier is adjusted so as to exceed the losses within the probe, continuous oscillations will be produced.

If the diaphragm of the probe is exposed to a noncompressible material that offers greater mechanical resistance to the motion of the diaphragm, the transfer of energy to the pickup coil decreases. This results in a decrease in the signal fed back into the amplifier and a corresponding decrease in the signal available from the output of the amplifier. The decreased signal triggers a voltage-sensitive network that controls the output relay.

Two-Sensor Ultrasonic System Applications

Figure 8–6 illustrates applications of the two-sensor ultrasonic system. Figure 8–6A represents bin level control. Figures 8–6B through 8–6D represent web

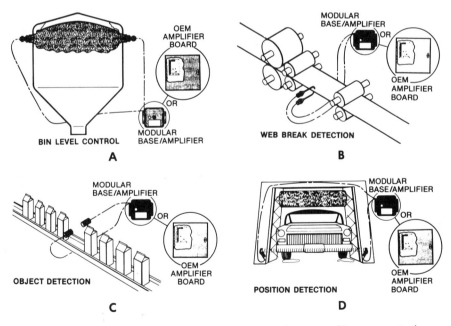

Figure 8–6 Two-Sensor Ultrasonic System Applications *(Courtesy, Delavan Electronics, Inc.)*

break, object, and position detection, respectively. In each application, two sensors are placed so that one acts as a receiver and one acts as a transmitter.

THE SINGLE–SENSOR ULTRASONIC WAVE CONTROL SYSTEM

The single-sensor ultrasonic sensor provides level detection control. The assembly consists of a stainless-steel sensor and a control amplifier (see figure 8–7). The sensor detects changes in the acoustic medium into which it transmits sound. The control amplifier furnishes the necessary electrical signals and a control relay to operate external functions.

The sensor is a small sealed unit. When liquid touches its face, a relay in the control amplifier operates, providing either on or off operation of the control system. The sensor operates ultrasonically and will sense any liquid.

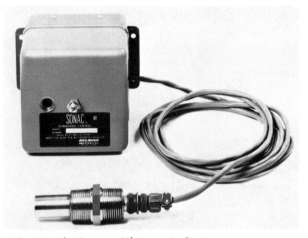

Figure 8–7 The Single-Sensor Ultrasonic System
(Courtesy, Delavan Electronics, Inc.)

The Single-Sensor Ultrasonic System Probe

The basic principle of single-probe operation is simple. The sensor has a moving diaphragm. When the diaphragm vibration is dampened by contact with liquid, the circuit will cease to oscillate and the control relay is changed. This operates the external function, which might be a level indicator or an overflow alarm.

Let's look in more detail at how the single ultrasonic probe assembly works. A simplified explanation of the probe operation involves the easy-to-understand fact that a moving diaphragm vibrates more freely in air than in liquid.

The probe consists basically of a diaphragm, a driving mechanism to produce motion in the diaphragm, a mechanical resonant system, and a means to detect motion of the diaphragm (see figure 8–8).

When energy is supplied to the driving mechanism, it causes the diaphragm to vibrate. The frequency is determined by the mechanical resonant system within the probe. The electrical energy will then be available in the motion detecting portion of the probe. The amplitude of the signal in the motion detecting coil will be proportional to the diaphragm motion.

Broadly stated, the driving section of the probe will be referred to as the *input* and the motion-detecting section of the probe referred to as the *output* of the probe. Connecting the output of the probe to the input of an amplifier and the output of the amplifier to the input of the probe forms a circuit. Any energy appearing in the output of the probe will be fed to the amplifier and there amplified. It will then be furnished to the input of the probe (see figure 8–9). This will cause vibrations to occur in the diaphragm and furnish a signal back to the amplifier for reamplification.

If the face of the probe is then exposed to a material that offers greater mechanical resistance to the motion of the diaphragm, the output circuit of the probe will sense this decreased motion. This results in a decrease in the signal fed back into the amplifier. The result is a corresponding decrease in the signal available from the output of the amplifier. If this additional loss in the circuit is in excess of the available adjusted gain in the control amplifier, the circuit will

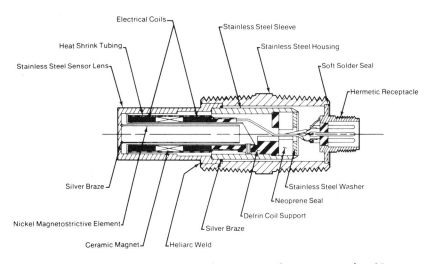

Figure 8–8   Cross-Section of a Single-Sensor Ultrasonic Probe *(Courtesy, Delavan Electronics, Inc.)*

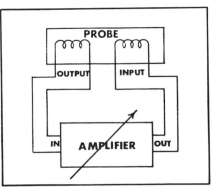

Figure 8-9 Simplified Schematic Single-Sensor Ultrasonic System *(Courtesy, Delavan Electronics, Inc.)*

then cease to oscillate. The control relay in the amplifier responds to the condition of oscillation or nonoscillation of the circuit.

The probe is basically a magnetorestrictive device consisting of a diaphragm, nickel tube, drive coil, and pickup coil.

The active area of the probe is the diaphragm. The center of the diaphragm is the point of greatest sensitivity to the environment. The effect of air pressure or gas pressure on the operation of the probe is negligible. Because the compressed gas is still compressible to a further degree, the net change in probe acoustic characteristics in relationship to air at normal atmospheric pressure is extremely small. Within the pressure rating limitations, the effect of liquid under pressure is almost nonexistent. A liquid or air pressure exceeding the recommended operating pressure limit on the sensor will cause a bowing or deformation of the diaphragm section of the probe. It will be sufficient to interfere with the mechanical structure of the internal assembly. At low pressure, this will overload the sensor and will fail to operate properly, but no permanent damage to the probe will result. At somewhat greater pressure, overloads will permanently deform the diaphragm and will ruin the probe.

Single-Probe Installation

The probe is designed for insertion through a bin wall. The probe is mounted either on top or bottom of a vessel. The outside or external portion of the probe contains an exit for the cable carrying the electrical currents to the probe (see figure 8-10).

The sensor is mounted in the horizontal position. If the sensor must be mounted vertically (from top of tank), it should be installed at an angle (30°) to

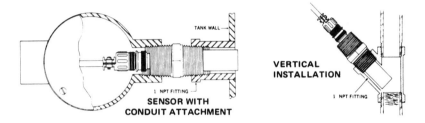

Figure 8-10 Ultrasonic Sensor Installation *(Courtesy, Delavan Electronics, Inc.)*

prevent entrapment of air bubbles on the sensor face. Air bubbles clinging to the sensor face could cause false responses.

The sensor face should be at least 1 inch away from nonliquid surfaces. The sensor should be installed so that the rear of the sensor, where the cable attaches, is kept dry.

Single-Sensor Ultrasonic System Applications

Figure 8–11 illustrates two applications of the single-sensor ultrasonic system. The lower application has the sensor installed in an overflow pipe. The upper

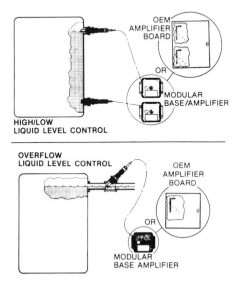

Figure 8-11 Single-Sensor Ultrasonic Systems in Application *(Courtesy, Delavan Electronics, Inc.)*

application monitors the upper level and the lower system with two separate ultrasonic systems.

MICROWAVE DETECTION

The state of the art in industrial controls has evolved from proximity devices such as inductive, capacitive, and magnetic devices through ultrasonic (electromagnetic frequencies above hearing level) into the world of microwave. The measurement parameter evolves around the microwave path attenuation and/or reflection.

PRINCIPLES OF OPERATION

Microwave Sensor Systems Defined

Microwave sensor systems are single-point noncontact controls that are designed to use low-energy-level microwaves to detect the level or position of liquids, bulk material, or solid objects.

Safety and Government Regulations

As you may know, the United States Department of Health & Human Services carefully monitors the manufacture, use, and sales of frequency devices. The FCC issues licenses or permits for purchase or operation of microwave devices (FCC Regulations, Part 15).

The maximum allowable power level, according to HEW regulations, is $10mW/cm^2$. The power level of the typical microwave industrial level control is $1.24mW/cm^2$ at the antenna. This is well within the requirements of regulations.

Microwave (nonionizing) radiation is that radiation which does not possess enough energy to cause displacement in atomic structures. That is, it cannot change atoms to ions. The effects of microwave radiation even at low levels, however, may cause minor heating. This heat in a typical microwave level detector system would represent less than 0.014 BTU/hour. Therefore heating effect is really not a concern.

System Description (See figure 8-12)

The block diagram of a typical microwave level control is shown in the figure. The system operates on the parameter of microwave path attenuation/reflection. The system, used to monitor these parameters, consists of a transmitter, a

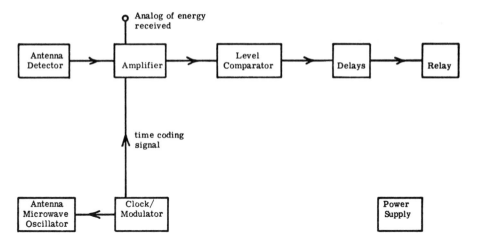

Figure 8-12 A Microwave Level Control System *(Courtesy, Delavan Electronics, Inc.)*

receiver, and a control circuitry. The measurement of the microwave path characteristics indicates the presence, absence, and/or the variation of materials in its path.

System Operation

When an object to be detected enters the microwave field between the transmitter and receiver, it either absorbs or reflects enough energy to cause a significant change in the signal level at the receiver. This change can be detected and used to initiate the desired output function.

The attenuation/reflection characteristics are governed by the electromagnetic energy propagation laws. Each time propagating microwave energy arrives at a material interface, some of the energy is reflected and some is transmitted into the new material. The amount of energy that travels in each direction is a function of the angle of incidence, polarization, and wavelength of the energy and the electrical properties—i.e., conductivity of permittivity and permeability of the materials involved. Attenuation that occurs in each region is a function of the electrical properties of that region and the wavelength of the energy involved.

There are three electrical properties that classify materials. The conductivity of the material is the reciprocal of the electrical resistance the material has. High conductivity means low resistance. One might be inclined to think that a low-resistance material would offer a low resistance to microwave energy, when the opposite is actually the case. For example, copper has a high conductivity (low resistance); microwave energy actually reflects off copper, and does not pass through.

The *permittivity* of a material is a measure of how much electrical energy a material can store.

The *permeability* of a material is a measure of how readily a material can be magnetized. Steel, of course, has high permeability. Dielectrics such as air, plastics, and nonferrous metals all have a relative permeability of one. Since microwave energy cannot "look through" ferrous and nonferrous metals owing to the conductivity, permeability is not considered a variable to consider in a microwave control system.

Application in Microwave Detection

Since microwaves are reflected by metals, and most process vessels are made of metal, a "window" transparent to the microwaves needs to be installed at the desired control point in the bin or tank. The window material is selected on the basis of its suitability to the process material handled, and its microwave transmission characteristics. Materials that transmit microwaves with minimal loss include many ceramics, glass, and plastic compounds that possess excellent resistance to impact, abrasive wear, temperature, pressure, and/or chemical corrosion. It should be noted that within these general categories, some materials (owing to their specific composition) are poor conductors of microwave, so care should be taken in the selection of this item.

Level Control of Liquids and Solids (See figure 8–13)

Level control of liquids and solids in portable bins and tanks can use this control technique. Formerly, to detect level in a portable bin a device was inserted into the bin, or the bin was placed on a scale and level was inferred from the weight.

Using microwaves, a control unit is permanently mounted at the desired control point at the loading station. The bin, if constructed of plastic or wood, normally requires no modification (see figure 8–13A). If it is made of metal, windows (typically made of UHMW polyethylene) must be installed at the desired height (see figures 8–13B and C). When the bin is positioned at the loading station, a microwave path is established through the bin until the material reaches the control level and absorbs or reflects the microwaves, breaking the path and initiating the desired output function.

The reliability of this technique depends on several factors. One of the most important factors is that the product must possess adequate attenuation for reliable detection, yet the attenuation should not be so great that normal

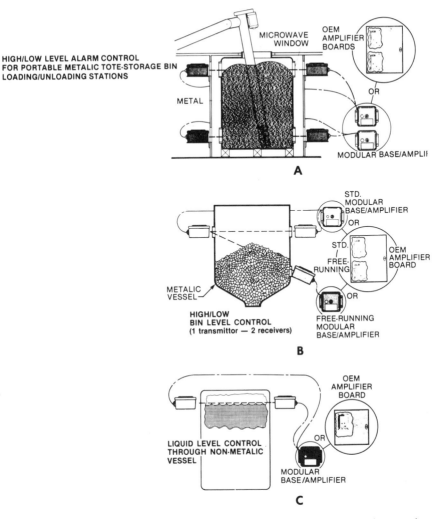

Figure 8–13 Level Control of Liquids and Solids *(Courtesy, Delavan Electronics, Inc.)*

product buildup on the sides of the container exceeds the dynamic range of the device.

High-Temperature Material Level Control (See figure 8–14)

Since microwaves are unaffected by thermal gradients, they are used successfully to detect the level of combustible material or the position of objects in furnaces. Since it is usually desirable to beam the microwaves through the refrac-

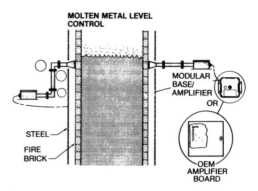

Figure 8-14 High-Temperature Material Level Control *(Courtesy, Delavan Electronics, Inc.)*

tory (heat-resistant material used to line furnaces) rather than physically penetrate it, care must be taken in the selection of the refractory material. Some refractory products exhibit minimal loss to microwaves at operating temperatures, while others can present extreme attenuation to the source.

Plugged Fill Chute Control (See figure 8-15)

Microwave control offers some distinct advantages on plugged chute detection of abrasive or physically abusive products. In addition, it can overlook buildups of many materials that tend to coat and foul other contact-type

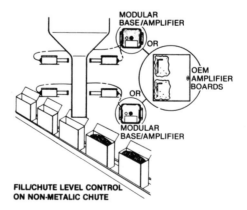

Figure 8-15 Plugged Fill Chute Control *(Courtesy, Delavan Electronics, Inc.)*

devices. Since it is a noncontact device, the control is separated from the process by the window material. Plastic materials exhibit exceptional resistance to impact and abrasion within its temperature limits, and its slippery surface resists product buildups.

Disguised-Object Detection (See figure 8–16)

Since microwaves pass through plastics, glass, cardboard, dry wood, etc. with relative ease, they can be used to detect the level, quantity, or presence of various objects contained within a sealed container. For example, the level of perfume within an opaque glass bottle can be checked; the position of the plate stack and the electrolyte level of the individual cells of a battery can be verified; the count of food sticks in a package confirmed where net weight is an unacceptable check.

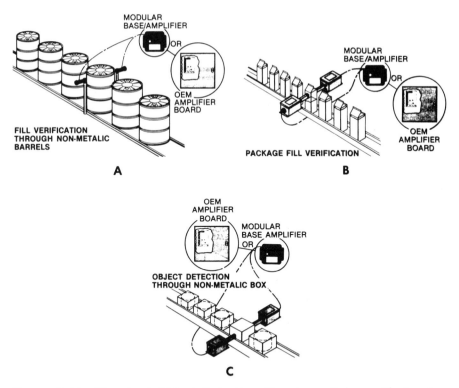

Figure 8–16 Disguised Object Detection (*Courtesy, Delavan Electronics, Inc.*)

In figure 8–16A, a microwave receiver and transmitter are used to detect the level of milk in cartons. In figure 8–16B, they are used to detect liquid or solid level in nonmetallic barrels.

In figure 8–16C, microwave is used to ensure that nonmetallic boxes were indeed filled with an object.

The Advantages of Microwave Control

The technique is noncontact. The process material cannot damage the control, the control cannot contaminate the process, and it can be serviced without process downtime.

Microwaves are unaffected by heat, dust, noise, fog, smoke, and can overlook many process buildups that collect on the window.

Low-level microwave energy is safe nonionizing radiation. Power levels are well below established safe standards. There are no licensing requirements for the purchaser in the United States or Canada.

A TYPICAL MICROWAVE DETECTION SYSTEM (See figure 8–17)

Typical of the microwave detection systems is the Delavan modular base/amplifier Model M70 with long-range transmitting/receiving sensors, Model MT80.

The modular base/amplifier is installed in a NEMA 1 drawn-steel enclosure.

The two transmitting/receiver sensors are long-range transmitting/receiv-

Figure 8–17 A Typical Microwave Detection System *(Courtesy, Delavan Electronics, Inc.)*

ing sensors with 10dB-gain horn antennas. Maximum range is 120 feet in air. Temperature range is $-20°F$ to $+140°F$ ($-30°C$ to $+60°C$). Each sensor, along with its 10dB-gain horn antenna, is installed in a NEMA 4 drawn-steel enclosure.

REFERENCES

Technical data and illustration was supplied by a state-of-the-art manufacturer. Permission to reprint was given by the following company:

Microwave detection and ultrasonic sensing—Delavan Electronics, Inc., Scottsdale, Arizona
All copyrights © are reserved.

9
Photodetection

The sensing of light is called photodetection. Photodetection is usually related to three phenomena. These are photoemissive, photoconductive, and photovoltaic actions. All quantum photodetectors respond directly to the action of incident light. The first, *photoemission,* involves incident light that frees electrons from a detector's surface. This usually occurs in a vacuum tube. With *photoconduction,* the incident light on a photosensitive material causes the material to alter its conduction. The third, *photovoltaic action,* generates a voltage when light strikes the sensitive material of the photodetector.

State-of-the-art light sensing is accomplished using solid-state devices. Therefore, we shall limit this chapter to the discussion of photoconductive and photovoltaic action.

The sun and stars are the natural sources of light. Reflective sources may include the planets and moons. The amount of light energy received from the sun is dependent on the distance the receiver is away from the sun. The amount of light energy (solar irradiance) received from the sun is $140mW/cm^2$ outside the earth's atmosphere. This is defined as the *solar constant.* The amount of solar irradiance that reaches the earth is around $100mW/cm^2$ if the day is clear and cloudless.

Light travels in waves similar to the electrical sine wave. Light waves are also considered to be part of the electromagnetic frequency spectrum. Light wavelengths are in length between 0.005 and 4000 microns.

The term "light" has been altered by technical advances to include wavelengths other than those that are visible. The new term "light" deals with wavelengths visible and not visible to the human eye. These wavelengths are called the *optical frequency spectrum.*

236

There are three basic levels of the optical spectrum. These are:

1. Infrared—Band of light wavelengths which are too long for response by the human eye
2. Visible—Band of light wavelengths to which the human eye responds
3. Ultraviolet—Band of light wavelengths which are too short for response by the human eye

ELECTROMAGNETIC WAVES

Light energy, as other forms of energy, is fairly predictable. Light energy travels in waves similar to the electromagnetic waves of electricity. The waves in electricity are called *sinusoidal* or *sine waves*. The sine wave has an angular rotation of 360 degrees. It is maximum at 90 degrees and 270 degrees, and minimum at 0, 180, and 360 degrees. See figure 9–1.

The sine-wave phenomenon is the same in light waves as in electrical waves. Each wave is characterized by four basic quantities: velocity, amplitude, frequency, and wavelength. Each wave has a positive and a negative alternation. Each light wave has a time period in which it goes through its complete cycle or alternation. Some waves take a shorter time to complete the cycle. For instance, from a home electrical socket, 60 cycles per second current is taken. This means that in one second the cycle repeats 60 times. The time for one cycle to take place is 1/60 of a second or 16.6 milliseconds.

$$T = \frac{1}{F}$$

where F = frequency in hertz
T = time in seconds

This formula can be converted to calculate frequency when time is known.

$$F = \frac{1}{T}$$

The time for one cycle to take place is called a *period* and is measured in seconds. The number of cycles that takes place within a time period is called a *hertz* or cycles per second.

The physical length in space of the wave is called its *wavelength*. It is this entity that concerns us in optoelectronics. All electromagnetic waves travel at the same velocity—that is, approximately 300,000,000 meters per second or

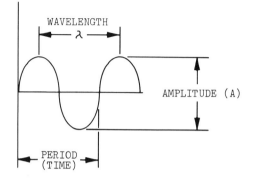

Figure 9-1   The Electromagnetic Wave

186,000 miles per second. The amplitude of the wave is the magnitude of the wave vibration. The frequency of the wave is the number of waves passing a point in a second. The wavelength is measured in meters per second, and is noted as lambda (λ). The wavelength is the distance between two consecutive wave peaks. Calculations of wavelength are as follows:

$$\lambda = \frac{V}{f}$$

where λ = wavelength in meters per second

V = velocity of light in meters

f = frequency of the wave in hertz
(cycles per second)

LIGHT RECEPTION

The reception of light is completely dependent on what use or application is intended. If the desire is to light a room, the light should reflect diffusely from most objects in the room, with absorbtion dependent on the materials and colors in the room. There would be little or no use for refractive light, which is light that is absorbed. If the light were to be used for photodetection, the amount of absorption would be extremely high (100 percent if possible). The amount of reflection would be held to a minimum. The application then decides the use of materials that are low or high absorbers, reflect or do not reflect, and refract or do not refract.

Surface Reflection (See figure 9-2)

Whenever a beam of light from one medium, such as air, strikes a second mirrorlike medium such as glass, part of the beam is reflected and the other part is

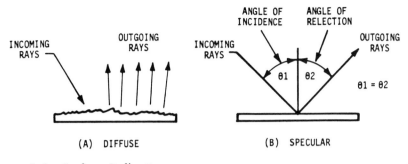

Figure 9-2 Surface Reflection

refracted. The angle at which the incident rays strike the second medium is called the *angle of incidence.* This angle is the angle made by the incident beam and a line normal (perpendicular) to the boundary of the two media. Part of the incident beam is reflected at an angle called the *angle of reflection.* This angle is made by the reflected beam and a line normal (perpendicular) to the boundary of the two media:

Angle of incidence = Angle of reflection

If the angle of incidence varies, so does the angle of reflection by the same amount. The angle of incidence, the angle of reflection, and the angle of refraction all lie in the same plane.

Surface Refraction (See figure 9-3)

Whenever a beam of light passes from one medium, such as water, the beam separates at the intersection. Part of the beam is reflected back into the incident medium (air) and part is refracted into the second medium (water). The angle made by the incident beam and a line normal (perpendicular) to the intersection is called the *angle of incidence.* The angle made by the reflected beam and the line normal to the intersection is called the *angle of reflection.* The angle of incidence and the angle of reflection are equal. The angle made by the refracted beam and a line normal (perpendicular) to the intersection is called the *angle of refraction.*

The ratio of the indexes of refraction of the two indexes of refraction is thus:

$$n = \frac{n2}{n1}$$

or n = index of refraction

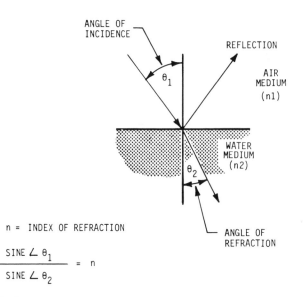

Figure 9-3   Refraction

The ratio of the sine of the angle of incidence to the sine of the angle of refraction is equal to the index of refraction.

$$n = \frac{\text{sine} < \theta\,1}{\text{sine} < \theta\,2}$$

therefore:

$$\frac{n2}{n1} = \frac{\text{sine} < \theta\,1}{\text{sine} < \theta\,2}$$

or $n1$ sine $< \theta\,1 = n2$ sine $< \theta\,2$

Furthermore, the ratio is a constant ratio. Whenever the angle of incidence changes, the angle of refraction changes to retain the ratio. This is Snell's Law. Snell was a Dutch astronomer and professor of mathematics at the University of Leyden in Holland.

Absorption

When a beam of light enters matter, its intensity decreases as it travels further into that medium. This is called *absorption*. There are two types of absorption—general and selective. General absorption is said to reduce all wavelengths of the light by the same amount. There are no substances that absorb all wavelengths equally. Some material, such as lampblack, absorbs nearly

100 percent of light-ray wavelengths. With selective absorption, the material selectively absorbs certain wavelengths and rejects others. Almost all colored things such as flowers and leaves owe their color to selective absorption. Light rays penetrate the surface of the substance. Selective wavelengths are absorbed while others are reflected or scattered and escape from the surface. These wavelengths appear in color to the human eye.

Scattering

The term "scattering" is differentiated from "absorption" in the following manner. With true absorption, the intensity of the beam is decreased in calculable terms as it penetrates the medium. Light energy absorbed in the material is converted to heat motion of molecules. Consider a long tunnel where you can only see light from one end. As you walk nearer that end, the light gets brighter. The light is absorbed as it travels through the tunnel at the absorption rate of the air medium. If the tunnel is then filled with a cloud of smoke, the smoke would scatter some of the light from the main beams. Therefore, the intensity of the light from a fixed distance will decrease. You may observe scattering effects by watching shiny dust particles as the sun shines through a window. Parts of the rays are scattered by the dust particles.

A scattering type known as *Rayleigh scattering* is caused by microirregularities in the medium. A wavelength passing through the medium strikes these microirregularities in the mainstream of the waves. The waves reflected from the microparticles are spherical and do not follow the main wavelength, but scatter. Therefore the intensity of the beam is diminished.

Atom Energy States (See figure 9–4)

Each atom has several energy states. The lowest energy state is the *ground state*. Any energy state above ground represents an *excited state* and can be represented by terms such as *first-* or *second-energy states,* or *intermediate-energy states.* If an atom is in one of its higher energy states and decays to a lower energy state, the loss of energy (in electron volts) is emitted as a *photon.* The energy of the photon is equal to the difference between the energy of the two energy states. The process of decay from one energy state to another is called *spontaneous decay* or spontaneous emission.

An atom may be irradiated by some light source whose energy is equal to the difference between ground state and some energy state. This can cause an electron to change from one energy state to another by absorbing source-light energy. This transition from one energy level to another is called *absorption.* In

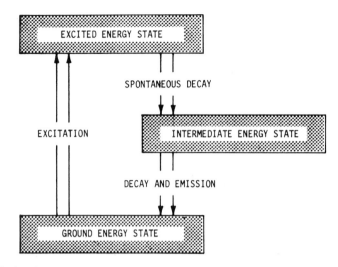

Figure 9-4 Atomic Energy States

making the transition, the atom absorbs a packet of energy called the photon. This is similar to the process of emission.

The energy absorbed or emitted (photon) is equal to the difference between the energy states that have been transitioned. For instance, $E_2 - E_1 = E_p$ (energy of photon):

$$E_2 - E_1 = E_p \text{ (energy of photon)}$$
$$\text{also } E_p = hf$$
$$\text{where } h = \text{Planck's constant}$$
$$\text{and } f = \text{frequency of the radiation}$$

The value of h has a dimension of energy and time. The value is 6.625×10^{-34} joule-second. The value of f is in cycles per second (hertz). Photon energy is said to be directly proportional to frequency.

Photon energy may also be expressed in terms of wavelength. You may recall that wavelength is equal to the speed of light in meters per second divided by frequency:

$$\lambda = \frac{\nu}{f}$$

$$\text{where } \lambda = \text{wavelength}$$
$$\nu = \text{speed of light in meters per second}$$
$$f = \text{frequency of the electromagnetic wave in}$$
$$\text{cycles per second (hertz)}$$

The formula can also be stated in the following manner:

$$\nu = f\lambda$$
$$\text{and } f = \frac{\nu}{\lambda}$$

If we substitute this value of frequency (f) in the formula for the energy of a photon, we have a second formula which relates the energy of the photon to wavelength:

$$E_p = hf$$
$$E_p = h\left(\frac{\nu}{\lambda}\right)$$
$$E_p = \frac{h\nu}{\lambda}$$

You may recall the Bohr theory from other basic atomic studies. The theory described the atom as having a nucleus which consists of neutrons and protons. Electrons rotate around the nucleus in shells or rings. The outside ring holds valence electrons. These electrons are the furthest away from the nucleus and therefore are not held as tightly in place as others which are closer to the nucleus. If there is enough radiant energy applied to the atom, the electrons can be removed from their orbits. The electrons can then escape from the force applied by the nucleus and become free carriers. The amount of energy required to remove the electron from the valence rings (excite the valence electron into an excited state) is called the *energy gap*. Listed in table 9–1 are some of the solid-state materials used as photodetectors, along with their energy gaps.

TABLE 9–1. Photodetection Materials and Their Energy Gaps

Atom	Energy Gap in eV
Cadmium Sulfide (CdS)	2.4
Cadmium Selenide (CdSe)	1.7
Silicon (Si)	1.1
Germanium (Ge)	0.7
Lead Sulfide (PbS)	0.37

PHOTODETECTION

The human eye detects radiation in the wavelength range of 3900 to 7500 angstroms. Other detectors are capable of detecting electromagnetic radiation from radio waves to X-rays. The wavelength detection that we are considering includes infrared, visual, and ultraviolet wavelengths.

The path of these rays may be blocked completely by material which we call *opaque*. Material that allows part of the light to pass through is called *translucent*. Material called *transparent* allows all or most of the light to pass through.

A detector material is specially selected so that when it is exposed to light rays it will absorb the light energy. If the detector responds to the light energy and not the wavelength, the detector is said to be *nonselective*. If the detector responds by varying detection for different wavelengths, it is *selective*. Selectivity or responsivity is the detector's response per unit of light. Wavelength responsivity is called *spectral response*. Frequency response is the speed by which the detector responds to radiation. Fluctuations in output current and/or voltage are referred to as *noise*. Noise is usually caused by current that flows in the detector regardless of whether or not light is applied. Current, such as this, is termed *dark current*, for it flows even without radiation. A common specification for a detector is the signal-to-noise ratio. This ratio is the noise current divided by the signal current.

The most basic form of detection is the *photoresistor*. The photoresistor is a small slice of photoconductive material whose resistance decreases or increases as light energy is applied. Electrons are released by the light and flow toward a positive power supply. The basic task of the photoresistor is to convert light energy to electrical energy. The photoresistive material is nonreflective. There are no junctions in a bulk photoresistor.

A second detector is the *single-junction photodiode*. A photodiode is the optical version of the standard diode. It is constructed of a pn junction. Photons of light energy are absorbed into the device. Hole-electron pairs are generated. The pairs are combined at different depths within the diode depending on the energy level of the photon. A wide, thin surface area is used to ensure maximum absorption. Current flow is dependent on the amount of radiation that is absorbed.

Photodiodes operate with and without dc bias. The solar cell is a photodiode that is heavily doped. The depletion area is extremely thin, as is its radiation area. The cell is coated to avoid reflection. Hole-electron pairs diffuse to the depletion area of the diode, where they are drawn out as useful current. Output current is dependent on input radiation. Solar cells are not biased and are photovoltaic in operation.

Phototransistors are two-junction devices which have a large base area. The base region of the phototransistor absorbs the photons of energy and generates hole-electron pairs in the large base collector region. The collector, being reverse-biased, draws the holes toward the base and the electrons toward the collector. The forward-biased base emitter junction causes holes to flow from base to emitter and electrons to flow from emitter to base. Forward bias causes the phototransistor to operate just as the conventional transistor operates. The basic function, then, is that the light energy induces the transistor to conduct.

Photo-field-effect transistors are constructed in the same manner as the standard-junction FET. A lens is used to focus light on the gate. The gate, being light-sensitive, excites its electrons into a conduction band. Electron movement causes current flow and a voltage drop. The voltage drop induces the FET load current through its drain resistance and thus modifies the drain to source voltage.

Each of the devices and detection methods that we have discussed have applications in the infrared and the ultraviolet wavelengths with detector material changes.

Other detection devices include photoposition detectors, the photo-thyristor, and opto-isolators. *Photoposition detection devices* provide an output that is proportional to the direction of the input beam. The *photothyristor* is a four-layer pnpn device. Photons of energy create hole-electron pairs in a thin junction. These pairs are drawn across all three junctions, producing a current from anode to cathode. The *opto-isolator* combines a light source with a detector within a single package. Current is applied to a source LED (light-emitting diode).

The LED emits photons, which are detected by a detector. The detector can be in the form of a photoresistor, photodiode, phototransistor, or photothyristor.

Photoconduction (See figure 9-5)

Operation through photoconduction involves a change in resistance of the photosensitive material. A wafer of photoconductive material is placed

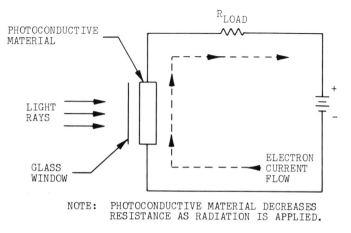

R_{LOAD}

PHOTOCONDUCTIVE
MATERIAL

LIGHT
RAYS

GLASS
WINDOW

ELECTRON
CURRENT
FLOW

NOTE: PHOTOCONDUCTIVE MATERIAL DECREASES
        RESISTANCE AS RADIATION IS APPLIED.

Figure 9-5 Photoconduction

underneath a glass window to protect it from exposure. The photoconductive material is tied to a load resistance and a power source. The clear glass window allows light radiation to strike the photoconductive material, freeing valence electrons. The resistance of the photoconductive material decreases, causing current through the load to increase. The resistance of the photoconductive material may change from several million ohms to several hundred, depending on the current demand of the device.

Photovoltaic-Cell Operation (See figure 9–6)

Operation of the photovoltaic cell involves the use of dissimilar metals to generate an electromotive force in response to radiated light. In figure 9–6, a light-sensitive material is placed beneath a thin layer of transparent metal and next to a dissimilar metal. The light-sensitive material is exposed to radiation through the thin transparent metal, which acts as a filter. When exposed, free electrons are removed from the light-sensitive material, causing electrons to flow to the dissimilar metal. This creates current flow and a difference of potential between the two terminals connected to the load.

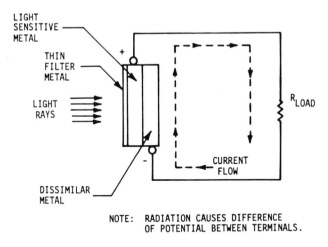

Figure 9–6 Photovoltaic Cell Operation

Solar Cells

The solar cell is probably the greatest user of the photodiodes in the photovoltaic mode. The solar cell converts the sun's light energy to electrical energy. In space, the solar cell is used to provide power for spacecraft. The efficiency of the cell at high altitudes is very high, because beyond the earth's atmosphere the sun's rays are not inhibited. Solar cells are being used on the earth for many applications, including the heating of homes.

Solar light has its peak at around 500 nanometers. Power from this light is near 100 milliwatts per centimeter if it can all be collected.

The solar-cell structure is very simple and similar to the photodiode. The depletion area is very narrow. The active area is very large to intercept the maximum radiant flux. Speed of the solar cell is very low. The diode is operated in the photovoltaic mode without bias. Photons of adequate energy create electron-hole pairs. The pairs are drawn from the narrow depletion area with a high probability of recombination. The current created from the recombination is directed through a very low resistance to obtain maximum power transfer.

Solar cells in space may reach 75 percent efficiency, while those on the earth are still struggling to obtain 20 percent. There have been indications of breakthrough in efficiency. Even a quantum jump of 5 percent would be significant.

Solar cells are especially useful in remote areas for recharging batteries to operate communication equipment. Solar cells help in operating weather devices.

Cell banks in remote areas must be set to obtain the most sunlight throughout the year. If servos are used to place the solar-cell banks in the right position, the current created by the cell is used up in moving the bank. Therefore, the mechanics of positioning from season to season, along with varying temperatures, cloudy days, and sunlight time, are certainly dominant obstacles against efficiency.

Photodetection Materials

A detector material is specially selected so that when it is exposed to light rays it will absorb the light energy. If the detector responds to the light energy (flux) and not the wavelength, the detector is said to be nonselective. If the detector responds by varying detection at different wavelengths, it is selective. Selectivity or responsivity is the detector's response per unit of light. Wavelength responsivity is called *spectral response.* Frequency response is the speed by which the detector responds to radiation. Fluctuations in output current and/or voltage are referred to as *noise.* Noise is usually caused by current that flows in the detector regardless of whether or not light is applied. Current such as this is termed *dark current,* for it flows even without radiation. A common specification for a detector is the signal-to-noise ratio. This ratio is the noise current divided by the signal current.

The Photoresistor

Photoresistor resistor materials are called *bulk photoconductors*. Bulk photo-conductors are usually made from cadmium sulfide (CdS), cadmium selenide (CdSe), lead sulfide (PbS), and silicon (Si). The type of material is dependent on the application. The reader will note that bulk photoconductors do not have a junction. They are simply made from one material. This material should provide a broad wavelength from ultraviolet through infrared. The ultimate purpose of the bulk photoconductor is to convert light intensity to current flow. Therefore, the material selected should free electrons easily with the application of light. Light should not reflect, but be transmitted easily into a detector. The detector should be transparent at the desired wavelength. To be efficient, the bulk (resistive) material should be extremely sensitive to light radiation and as easy to apply as a resistor. The primary problem with photoresistive material is that a problem arises with temperature stability for faster-acting materials.

The photodiode is usually made from silicon or germanium, much the same as for standard diodes.

The left drawing in figure 9–7 is a cross-section drawing of a conventional photodiode. There have been many processes by which pn junctions have been formed. The technique that has become the most popular and most utilized by industry is the *planar process*. This process is utilized by integrated-circuit manufacturers. The process uses silicon as the solid and the dopants as gases. By diffusion, the dopant gases penetrate the solid surface of the silicon. Diffusions may be made on diffusions; therefore several layers of dopants may be diffused on one device. This serves to provide manufacturing versatility.

The basic pn junction used in production of the photodiode is then the planar diffused. In the illustration, N-type bulk silicon is diffused on one side by N+ dopant and on the opposite side by P+ dopant. A depletion region between the N and P exists free of current carriers. It is in this depletion area that the photons should be absorbed. To operate in the photoconductive mode, the device must be reverse-biased. To operate photovoltaicly, no bias is required.

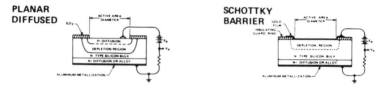

Figure 9–7 Photodetector Manufacturing Processes *(Courtesy, United Detector Technology, Inc.)*

An active area exposes the P + diffusion to light beams. The light beams are absorbed into the semiconductor. When photons of energy are absorbed into the material, electron-hole pairs are formed. Short, medium, and long wavelengths of photon power are absorbed at different depths within the pn junction. The depth is dependent on the photon wavelength. Short wavelengths, of course, are absorbed near the surface. Long wavelengths may penetrate the entire structure. To be most useful, the wavelengths should be absorbed in the depletion area.

Current is produced by electron-hole pairs being separated and drawn out in directions of more positive or negative sources, whichever is the case. If the electron-hole pairs happen outside the depletion area, they will usually combine and no current will be produced. The active region (P + diffusion) should be extremely thin to ensure maximum penetration. As in other reverse-biased diodes, the depletion area can be made larger by increasing the reverse bias.

In the right drawing of figure 9–7, the *Schottky barrier photodiode* is illustrated. Operation of the Schottky photodiode is much the same as the conventional pn junction. The Schottky diode (often called a surface diode) differs in the method by which the p-type material is formed. In the conventional photodiode, the method is diffusion. In the Schottky photodiode, the method is metallization. In the case illustrated, the active region has a thin gold film metallized to the N-type silicon bulk. The Schottky barrier photodiode, has some advantages over the conventional photodiode. It operates well at wavelengths less than 500 nanometers and has much simpler fabrication processes. The Schottky photodiode does not operate well at high temperatures or with high light power.

The PIN Photodiode

The PIN diode is so called because of the layer material by which it is constructed. The word PIN is an acronym for P-type, intrinsic, N-type materials. A PIN photodiode is one in which a heavily doped P region and a heavily doped N region are separated by a lightly doped I region. In figure 9–7, the N-type silicon bulk would represent the I region. The resistance of the I region can range from 10 ohms per centimeter to 100,000 ohms per centimeter. The P and N regions are less than one ohm per centimeter. Since a depletion area can extend further into a nondoped or lightly doped region, the PIN photodiode has an extremely large depletion area. This large depletion area provides the PIN photodiode with much faster speeds, lower noise, and greater efficiency at longer wavelengths.

The Phototransistor

The phototransistor is made from silicon and germanium and has much the same characteristics as the standard or conventional transistor. Phototransistors are constructed in much the same manner as photodiodes and standard transistors, using the planar diffused method. The base area of the phototransistor is usually made large so as to provide an area which incident light can penetrate and generate electron-hole pairs. Phototransistors are subject to the typical problem of all transistors—temperature variations. This problem may be solved with biasing techniques and thermal stability resistors.

Infrared Detectors

Infrared detectors must use thermal detectors because their wavelengths are longer. Lead sulfide (PbS) responds to infrared wavelengths of 2 to 4μm. Indium antimonide (InSb) is sensitive to infrared wavelengths up to 5.6μm. Silicon (Si) and germanium (Ge) are only sensitive to the near-infrared wavelengths.

Ultraviolet Detectors

Ultraviolet detectors utilize materials such as telluride cesium or a fluorescent coating of sodium salicylate as cathodes within phototubes. Bulk semiconductors made of cadmium selenide (CdSe) may respond to ultraviolet in the range 300 to 400 nm.

PHOTODETECTOR APPLICATIONS

Photodetectors are extremely versatile in that they may be used with a great variety of electronic devices to perform an equally great variety of tasks.

In figure 9–8A, a bulk photoresistor is placed on the input leg to an operational amplifier. The light intensity varies the resistance, which varies the input to the amplifier. R_F is a feedback resistor, while R1 sets the operating level of the operational amplifier.

Figure 9–8B illustrates a pn-junction solar cell in the input leg of an operational amplifier. As in the photoresistor operation, the intensity of the light causes a current change to the operational amplifier.

In figure 9–9A, a photodetector transistor is used to drive a relay coil. Light beams, applied to the base, forward-bias the transistor. The transistor

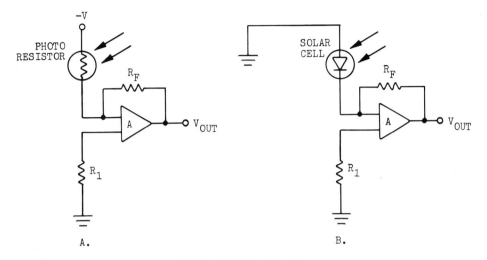

Figure 9-8   Photoresistor and Solar-Cell Applications

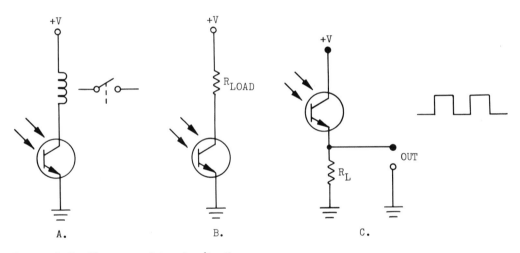

Figure 9-9   Phototransistor Applications

turns on, energizing the coil of the relay. Relay contacts close and remain closed until the light beam is removed from the transistor base.

In figure 9-9B, a phototransistor is used to drive current through a load. Light beams applied to the base forward-bias the transistor. The transistor turns on, causing current to flow through a load resistance.

In figure 9-9C, a phototransistor is used as a square-wave switch. This is used when light is applied in the form of bursts.

Photodiode Applications (See figure 9-10)

Photodiodes are used in photoconductive and photovoltaic applications.

In the photoconductive mode, the device must be operated with reverse-bias. The problem with reverse-bias is that the device develops reverse current, called dark current. The dark current is present with reverse-bias when no outside radiation is applied.

In the photovoltaic mode, the device operates with zero bias. This makes sense, since the photodiode in this mode is the generator producing the load voltage.

The photoconductive circuit has reverse-bias. Signal current is felt through the load resistor R_L in the photoconductive circuit. An operational amplifier is used in the photovoltaic circuit to amplify the signal felt by the photodiode. Since the input resistance of the operational amplifier is extremely high, the gain is essentially that of the operational amplifier. The dynamic resistance R_f is used for feedback.

Photodiodes that are manufactured to operate in the photovoltaic mode do so with zero bias. This is ideal for the elimination of black current in applications that demand this requirement.

Photodiodes that are manufactured for photovoltaic modes are used in radiometers, photometers, spectrometers, densitometers, colorimeters, particle

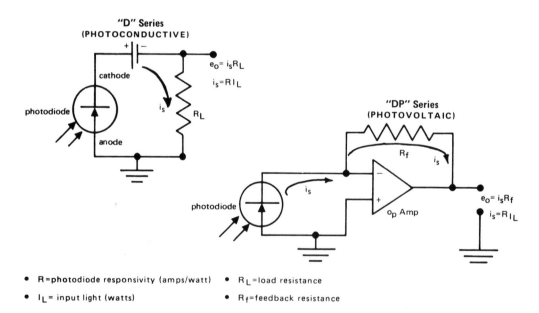

- R = photodiode responsivity (amps/watt)
- I_L = input light (watts)
- R_L = load resistance
- R_f = feedback resistance

Figure 9-10 Photodiode Applications *(Courtesy, United Detector Technology, Inc.)*

counters, in fluorescent analysis, replacement of phototransistors, point of scale scanners, intrusion alarms, opto couplers, E-O blood analyzers, and video disc recorders.

Large-area photovoltaic photodiodes have application in solar cells, food processing, illumination control, proximity detection, and automatic conveying.

The Schottky barrier photodiode represents the state of the art in large-area, high-sensitivity, fast-response, expanded spectral-range detectors. The cold-formed Schottky barrier preserves the high resistivity of the intrinsic (I) region of the PIN diode. This serves to provide low barrier capacitance, low dark current, and low noise. The thin gold contact used in metallization allows ultraviolet and blue light to pass easily into the depletion area where electron pairs are more effectively collected.

Schottky barrier photodiodes are used in radiometry, photometry, densitometers, colorimeters, laser power meters, laser range finders, ultraviolet flame detection, character recognition, alignment, and other position-sensing detection.

State-of-the-Art Photodetector Applications (See figures 9–11 and 9–12)

The state of the art use of photodetectors is with the fiber optic industry. A typical fiberoptic system is illustrated in figure 9–11. Operation of the system is

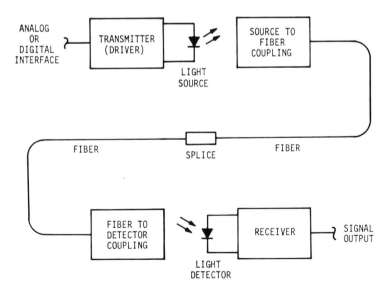

Figure 9–11 Simplified Fiberoptic System

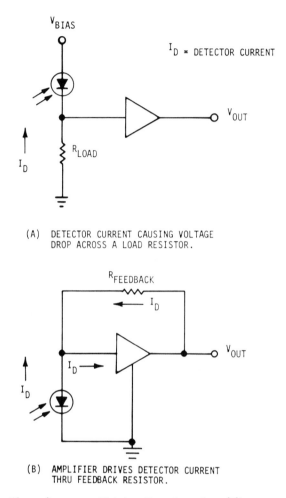

(A) DETECTOR CURRENT CAUSING VOLTAGE
     DROP ACROSS A LOAD RESISTOR.

(B) AMPLIFIER DRIVES DETECTOR CURRENT
     THRU FEEDBACK RESISTOR.

Figure 9–12   Photodetectors Driving Receiver Amplifiers

described as follows. A transmitter accepts an electrical signal and converts it to a current to drive a light source. The light source is usually an injection laser diode (ILD) or a light-emitting diode (LED). The light source launches (transmits) the optical signal into an optical fiber. The optical fiber provides a path for the light signal. The signal is reflected from side to side against the walls of the fiber. A photodetector such as a PIN diode detects the light signal and converts it to an electrical current. The receiver produces a low-noise, large-voltage-gain output from the detected signal which, it is hoped, is a replica of the input signal.

The function of the receiver is to accept low-level power from the detector

PIN diode and convert it into a high-voltage output. There are at least two methods of accomplishing this (see figure 9–12). In figure 9–12A, detector current produces a voltage drop across a load resistance. The voltage drop is amplified. An output voltage proportional to the transmitted signal is the result. In figure 9–12B, the amplifier output voltage is the effect of the amplifier driving detector current through the feedback resistance. Again, the output voltage is proportional to the transmitted signal.

State-of-the-Art Photodetector Device

Figure 9–13 is representative of state-of-the-art photodetectors. These particular detectors are dual-axis position sensors. The devices sense the centroid of the light spot and provide continuous analog output as the light moves to the limit of the active area. These detectors are planar-diffused PIN photodiodes. The detectors are the Models PIN-SC/4D and PIN-SC/10D, manufactured by United Detector Technology.

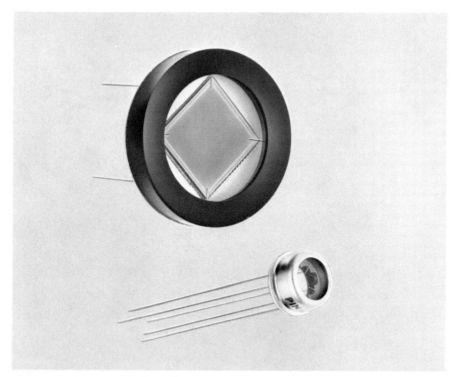

Figure 9–13 Typical Photodetector *(Courtesy, United Detector Technology, Inc.)*

REFERENCES

Reference data and illustrations were supplied by state-of-the-art manufacturers.

Permission to reprint was given by the following company:

Photodetectors—United Detector Technology, Santa Monica, California

All copyrights © are reserved.

10
Temperature Sensors

The sensing of temperature is usually accomplished with the aid of a thermoelectric device such as the thermocouple, the resistance temperature detector (RTD), or the thermistor. A thermoelectrical device is one that converts heat (temperature) to a corresponding electrical current flow or voltage level.

This chapter deals with these temperature sensors and their electrical properties.

TEMPERATURE SENSING

In all of science, heat is dissipated when energy is being converted or when some type of work is being done. In some cases, such as conversion of coal to steam to electric, large losses are caused by heat dissipation. In other adaptations, such as air conditioning, it is indeed desirable for heat to be removed. Whichever the case, the temperature transducer plays an important role in monitoring how much heat is present at specific points within the system.

Temperature Scales

Temperature is monitored by several scales in degrees. The two most used of these scales are Fahrenheit (F) and Celsuis (C). The latter is often referred to as *centigrade*. These two scales are the ones the layperson uses in everyday life. The base unit of temperature within the System International (SI) is the unit of Kelvin (K). Kelvin is used in scientific work. The SI units represent the metric system. Each one of these scales has desirable features. Probably the best way to illustrate these desirable features is with an illustration. In table 10–1, major points of each scale are represented on a chart. You will note that the coldest

TABLE 10-1. Comparison of Temperature Scales

Event	Kelvin (K)	SCALE IN DEGREES Celsuis (C)	Fahrenheit (F)
Water boils	373.15	100	212
Body temperature	310.15	37	98.6
Room temperature	295.15	22	72
Water freezes	273.15	0.0	32
Absolute zero	0.0	−273.15	−459.67

temperature possible is 0 degrees Kelvin. This establishes the desirability of the Kelvin scale. You will also note that water freezes at 0 degrees Celsius and boils at 100 degrees Celsius.

The methods for conversion of Celsius to Fahrenheit and vice versa are as follows:

$$C° = 5/9 \ (F° - 32°)$$
$$F° = 9/5 \ (C°) + 32°$$

The Celsius scale was developed by Anders Celsius, a Swedish astronomer. The Fahrenheit scale was devised by Gabriel Daniel Fahrenheit, a German physicist.

The Seebeck Effect (See figure 10–1A)

Thomas J. Seebeck, as we discussed earlier, discovered the thermocouple. This is also called the Seebeck effect. Seebeck fused two metal wires together on both their ends. He then heated one of the junctions and found that electron current flowed from one wire to the other. In this case, electron flow was from the copper wire to the iron wire. The heated junction is called the hot junction, while the other is called the cold junction. In figure 10–1B, the cold junction is replaced with a voltmeter. The voltmeter provides a closed-circuit path and monitors the difference of potential across the heated junction.

The potential developed across the heated junction is the thermocouple potential. Its polarity and magnitude are dependent upon the type of material of the two dissimilar metals (see figure 10–1C and 10–1D). In figure 10–1C, the positive polarity is on the chromel side of the thermocouple, while the negative polarity is on the alumel side. In a second example in figure 10–1D, the positive polarity is on the iron while the negative polarity is on the constantan.

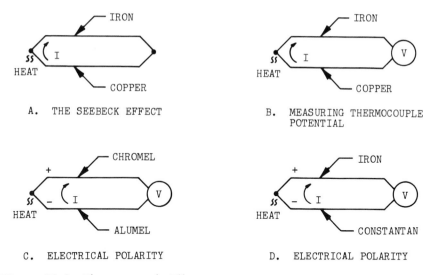

A. THE SEEBECK EFFECT

B. MEASURING THERMOCOUPLE POTENTIAL

C. ELECTRICAL POLARITY

D. ELECTRICAL POLARITY

Figure 10-1 Thermocouple Effects

The metal chromel is an alloy of nickel and chromium. Alumel is an alloy of nickel, magnesium, aluminum, and silicon. Constantan is an alloy of copper and nickel. We shall define the alloys more clearly later on in the chapter under the section "Thermocouple Materials."

The Peltier Effect (See figure 10-2A)

Jean Peltier, as we discovered earlier, applied current to a junction made by two dissimilar materials to develop what is now called the Peltier effect. In the illustration, an iron and a copper wire are fused together at their ends. A battery is placed in series with the iron lead. Current flows in the entire closed-loop circuit. As electrons flow from the iron material to the copper material, the junction (A) becomes hot. This is because the electrons are moving from the high-energy-state material of iron to a low-energy-state material of copper. The excess energy heats the junction. As electrons flow from the copper material to the iron material, the junction (B) becomes cold. This is because the electrons are moving from a low-energy-state material of copper to a high-energy-state material of iron. The thermal energy of the junction supplies the energy for transition. If the current were reversed, as in figure 10-2B, the junction (B) would become the hot junction, while junction (A) would become the cold junction. The phenomena would not change, however. The hot junction will always be where electrons are moving from a high-energy-state material to a lower-energy-state material.

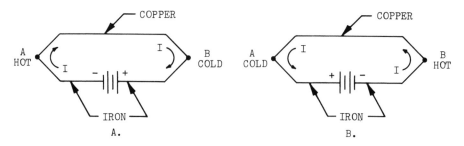

Figure 10-2   The Peltier Effect

Although the Peltier effect in application does not constitute a transducer or a sensor, it is appropriate to mention it as a *thermoelectric effect*. Its applications are with devices which require cooling or heating functions. The Peltier effect is applied with semiconductor materials such as bismuth telluride as a conductor of thermoelectric carriers.

The Faraday Effect (See figure 10-3)

Michael Faraday, an English scientist, found through experimentation that certain semiconductor materials decrease their resistance as temperature increases.

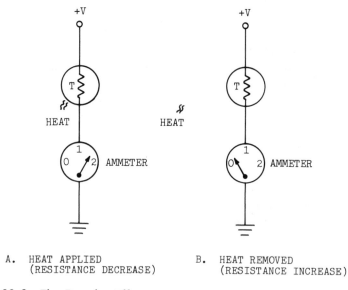

Figure 10-3   The Faraday Effect

The material is said to have a *negative temperature coefficient*. It was found later that oxides of cobalt, manganese, and nickel provide thermally sensitive resistance for temperature-involved applications. The resistance became known as the *thermistor*. In figure 10–3A, a thermistor is installed in a circuit with an ammeter. Heat is applied to the thermistor. The thermistor decreases resistance. Current flow increases and the ammeter reflects the increase. When heat is taken away, the resistance increases and current decreases (see figure 10–3B). The thermistor has, then, sensed the change in heat so that it can be monitored as current flow.

Resistance Temperature Detectors (RTD) (See figure 10–4)

As was previously discussed, the thermistor has a negative temperature coefficient. As temperature increases, the thermistor resistance decreases. Material such as platinum and nickel have positive temperature coefficients. These materials are known as *resistance temperature detectors*. In figure 10–4A, heat is applied to the RTD. The resistance of the RTD increases and current through the ammeter decreases. In figure 10–4B, heat is removed. The resistance of the RTD decreases and current through the ammeter increases. As with the thermistor, a meter can be calibrated to monitor a change in resistance or current as a result of a temperature change.

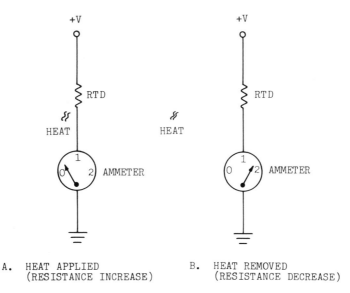

Figure 10–4 Resistance Temperature Detector (RTD) Operation

THERMOCOUPLES

Material Designations

Thermocouples are chosen for their ability to provide a uniform voltage-temperature relationship. When temperature changes, the thermocouple should produce a linear change in voltage output. Some thermocouples operate well at high temperatures; others operate best at low temperatures. Some are not subject to corrosion, humidity, or oxidation. Others may be contaminated by exposure to specific elements. The type of thermocouple chosen must meet the requirements of the job. Since this is the case, standards have been developed by industry and the National Bureau of Standards. Thermocouples are designated by letter types. The common metals (called base materials) are ANSI (American National Standard Institute) types T, E, J, and K. The more exotic metals used are called *noble metals*. These are more expensive but are able to operate at higher temperatures and have high resistance to oxidation and corrosion. These are ANSI types R and S. There are also types 1 through 5 (shown in figure 10–5) which are industry standards but are not ANSI symbols.

The designations do require some explanation, because ANSI also provides letter designations for alloy types. Table 10–2 supplies the ANSI letter designation for alloys versus the alloy trade name. Table 10–3 supplies the trade name and its manufacturer. Finally, table 10–4 lists the ANSI type and the metal alloy combinations used to manufacture the thermocouple. After the material there are (+) and (−) signs. The (+) polarity establishes the metal with the higher energy state.

TABLE 10–2. ANSI Designations Versus Trade Names of Alloys

ANSI Designation	*Alloy (Generic or Trade Name)*
JN, EN, or TN	Constantan, Cupron, Advance
JP	Iron
KN	Alumel, Nial T2, Thermokanthal KN
KP or EP	Chromel, tophel T1, Thermokanthal KP
RN or SN	Pure platinum
RP	Platinum 13% Rhodium
SP	Platinum 10% Rhodium
TP	Copper

TABLE 10-3. Alloys and Their Manufacturers

Alloy	Manufacturer
Advance T (Constantan)	Driver-Harris Co.
Alumel	Hoskins Mfg. Co.
Chromel	Hoskins Mfg. Co.
Cupron (Constantan)	William B. Driver Co.
Nial	William B. Driver Co.
Thermokanthal KP	The Kanthal Corp.
Thermokanthal KN	The Kanthal Corp.
Tophel	William B. Driver Co.

TABLE 10-4. ANSI Symbol and Its Thermocouple Alloys

ANSI Symbol	Thermocouple Alloy
T	Copper $^{(+)}$ versus constantan$^{(-)}$
E	Chromel$^{(+)}$ versus constantan$^{(-)}$
J	Iron$^{(+)}$ versus constantan$^{(-)}$
K	Chromel (+) versus alumel (−)
*G	Tungsten (+) versus tungsten 26% rhenium (−)
*C	Tungsten 5% rhenium (+) versus tungsten 26% rhenium (−)
R	Platinum (+) versus platinum 13% rhodium (−)
S	Platinum (+) versus platinum 10% rhodium (−)
B	Platinum 6% rhodium (+) versus platinum 30% rhodium (−)

* These letters are not ANSI symbols

Operating Ranges

The thermocouple does not generate an enormous amount of voltage (see figure 10-5). This is one reason for making a choice of the thermocouple depending on the temperature range of the particular need. A second factor in the choice of a thermocouple involves its ability to produce a voltage output. In figure 10-5, temperature versus millivolt curves are plotted for the ANSI-standard thermocouples. The reader will note that the base alloys (T, E, J, K) have millivolt outputs but operate in comparatively low temperature ranges. The noble alloys (R, S), however, operate in relatively high temperature ranges but have low voltage outputs. Finally the tungsten-rhenium alloys (3, 4) operate at

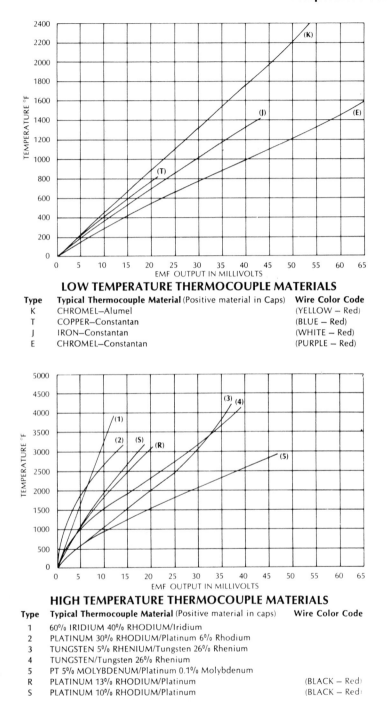

LOW TEMPERATURE THERMOCOUPLE MATERIALS

Type	Typical Thermocouple Material (Positive material in Caps)	Wire Color Code
K	CHROMEL–Alumel	(YELLOW – Red)
T	COPPER–Constantan	(BLUE – Red)
J	IRON–Constantan	(WHITE – Red)
E	CHROMEL–Constantan	(PURPLE – Red)

HIGH TEMPERATURE THERMOCOUPLE MATERIALS

Type	Typical Thermocouple Material (Positive material in caps)	Wire Color Code
1	60% IRIDIUM 40% RHODIUM/Iridium	
2	PLATINUM 30% RHODIUM/Platinum 6% Rhodium	
3	TUNGSTEN 5% RHENIUM/Tungsten 26% Rhenium	
4	TUNGSTEN/Tungsten 26% Rhenium	
5	PT 5% MOLYBDENUM/Platinum 0.1% Molybdenum	
R	PLATINUM 13% RHODIUM/Platinum	(BLACK – Red)
S	PLATINUM 10% RHODIUM/Platinum	(BLACK – Red)

Figure 10–5 Thermocouple Material Designations *(Courtesy, Hy-Cal Engineering)*

extremely high temperatures at an output voltage range between the base and the noble alloys.

Thermocouple Capabilities (See figure 10–5)

Each thermocouple has specific capabilities. These are as follows:

Iron constantan (ANSI symbol J)—The iron-constantan "J"–curve thermocouple with a positive iron wire and a negative constantan wire is recommended for reducing atmospheres. The operating range for this alloy combination is 1600°F for the largest wire sizes. Smaller size wires should operate in correspondingly lower temperatures.

Copper-constantan (ANSI symbol T)—The copper constantan "T"–curve thermocouple with a positive copper wire and a negative constantan wire is recommended for use in mildly oxidizing and reducing atmospheres up to 750°F. They are suitable for applications where moisture is present. This alloy is recommended for low-temperature work since the homogeneity of the component wires can be maintained better than other base-metal wires. Therefore, errors due to lack of homogeneity of wires in zones of temperature gradients are greatly reduced.

Chromel-alumel (ANSI symbol K)—The chromel-alumel "K"–curve thermocouple with a positive chromel wire and a negative alumel wire is recommended for use in clean oxidizing atmospheres. The operating range for this alloy is 2300°F for the largest wire sizes. Smaller wires should operate in correspondingly lower temperatures.

Chromel-constantan (ANSI symbol E)—The chromel-constantan thermocouple may be used for temperatures up to 1600°F in a vacuum or inert, mildly oxidizing, or reducing atmosphere. At subzero temperatures, the thermocouple is not subject to corrosion. This thermocouple has the highest emf output of any standard metallic thermocouple

Platinum-rhodium alloys (ANSI Symbol S, R)—Two types of "noble-metal" thermocouples are in common use; they are: (1) a positive wire of 90 percent platinum and 10 percent rhodium used with a negative wire of pure platinum; (2) a positive wire of 87 percent platinum and 13 percent rhodium used with a negative wire of pure platinum. These have a high resistance to oxidation and corrosion. However, hydrogen, carbon, and many metal vapors can contaminate a platinum-rhodium thermocouple. The recommended operating range for the platinum-rhodium alloys is 2800°F, although temperatures as high as 3270°F can be measured with the PT-30 percent Rh versus PT-6 percent Rh alloy combination (symbol 2).

Tungsten-rhenium alloys (symbol 3 and 4)—Two types of tungsten-rhenium thermocouples are in common use for measuring temperatures up to

4000°F. These alloys have inherently poor oxidation resistance and should be used in vacuum, hydrogen, or inert atmospheres.

Thermocouple Construction (See figure 10–6)

The thermocouple consists of two dissimilar metals fused together. The thermocouple assembly, as you can understand, involves a great deal more than two metals. The measuring junction is the hot junction. As was previously discussed, the hot junction is where electrons move from a high energy state to a lower energy state and release heat. The measuring junctions are specially constructed in a tube to provide the junction with support while still achieving uninhibited sensing of the environment to be measured. The supporting material is called a *sheath* and is made from metal such as inconel or stainless steel. The sheath is insulated from the junction with ceramic or magnesium oxide. There are three basic junction models. These are the exposed junction, the ungrounded junction, and the grounded junction.

Figure 10–6A illustrates an *exposed junction*. This junction is recommended for the measurement of static or flowing noncorrosive gas temperatures where response time must be minimal. The junction extends beyond the protective metallic sheath to give fast response. The sheath insulation is sealed at the point of entry to prevent penetration of moisture or gas.

Figure 10–6B shows an *underground junction*. This type is recommended for the measurement of static or flowing corrosive gas and liquid temperatures

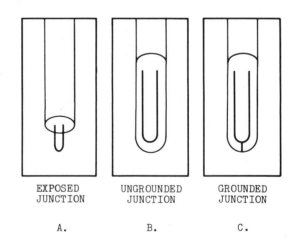

EXPOSED   UNGROUNDED   GROUNDED
JUNCTION   JUNCTION   JUNCTION

A.   B.   C.

Figure 10–6 Thermocouple Measurement Junctions (*Courtesy, Omega Engineering, Inc.*

in critical electric applications. The welded wire thermocouple is physically insulated from the thermocouple sheath by a hard, high-purity ceramic.

Figure 10–6C illustrates a *grounded junction*. This junction is recommended for the measurement of static or flowing corrosive gas and liquid temperatures and for high-pressure applications. The junction of this thermocouple is welded to the protective sheath, giving faster response than the ungrounded junction type.

The thermocouple measurement junction is, of course, attached to wires. The wires with junction are called *elements*. The wires are parallel, with their lengths varying depending on the job requirement. The ends of the wire are bent so as to fit neatly into terminal connections. Figure 10–7 illustrates a variety of these elements. This figure is self-explanatory.

A typical thermocouple assembly is illustrated in figure 10–8. The assembly consists of a connection head, extensions, and the thermowell. The lower end of the assembly is the thermowell. The thermowell encases the thermocouple element in a ceramic insulated sheath made of stainless steel. The sheath type is dependent upon the local environment. The probe type is dependent on the temperature range required.

The extensions are in the center of the probe. It is often desirable to extend the connection between the thermowell and the thermocouple connecting head for operational convenience or to avoid direct contact with hot surfaces. Also, it is often required that a union be employed in the extension to permit removal of the thermocouple element without twisting the leads.

The top of the thermocouple is called the *connection head*. The head provides protection for the elecrical terminations of the thermocouple and connec-

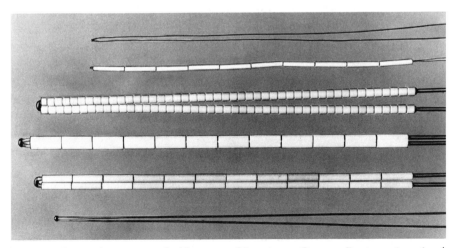

Figure 10–7 Thermocouple Elements *(Courtesy, Omega Engineering, Inc.)*

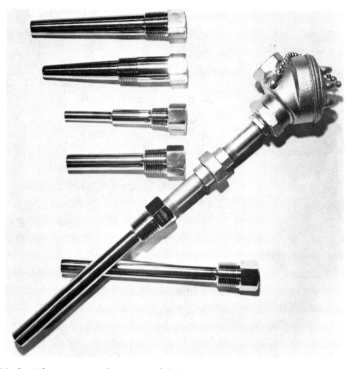

Figure 10–8   Thermocouple Assembly *(Courtesy, Omega Engineering, Inc.)*

tion to the associated instrumentation. This particular head is chained. It protects the electrical terminations with weatherproof gaskets. These thermocouples range from subminiature in size to over 100 feet long.

THERMOCOUPLE EXTENSION WIRES

In a large number of applications, the measuring junctions of a thermocouple system are located a considerable distance away from the reference junction and/or the indicating instrument. There are also the problems of junctions made by connecting wires to element connections. A further problem exists when one is using the same extension wire material as is used in the element. For instance, the use of platinum wire for extension compatibility may tend to be cost-prohibitive. The ratio in price between a platinum .020-inch diameter wire and an alumel .020-inch diameter wire is perhaps 100:1 per foot. There are special wires that can be purchased which are compensated to match platinum thermocouple elements. Wiring comes sheathed and unsheathed. It comes with

Figure 10-9 Typical Insulated Thermocouple Cable *(Courtesy, Omega Engineering, Inc.)*

Teflon coating and with insulation. Unsheathed wire is issued in matched pairs. Insulated wire comes in two or more strand cables. Cables must be small, durable, and easily formed. The cable must also be able to withstand high temperature and pressure.

The parameters that all extension wires must meet are the limits of error at specific temperature ranges.

Figure 10-9 illustrates a typical insulated thermocouple cable.

THERMOCOUPLE REFERENCE JUNCTIONS

Philosophy

When accurate thermocouple measurements are required, it is common practice to reference both legs to copper lead wire at the ice point so that copper leads may be connected to the emf readout instrument. This procedure avoids the generation of thermal emfs at the terminals of the readout instrument. Changes in reference-junction temperature influence the output signal and practical instruments must be provided with a means to cancel this potential source of error. It is accomplished by placing the reference junction in an ice-water bath at a constant $0°C$ ($32°F$). Because ice baths are often inconvenient to maintain and not always practical, several alternate methods are often employed.

Electrical Bridge Method

This method usually employs a self-compensating electrical bridge network, as shown in figure 10–10. This system incorporates a temperature-sensitive resistance element (R_T), which is in one leg of the bridge network and thermally integrated with the cold junction (T_2). The bridge is usually energized from a mercury battery or stable dc power source. The output voltage is proportional to the imbalance created between the preset equivalent reference temperature at (T_2) and the hot junction (T_1). In this system, the reference temperature of $0°C$ or $32°F$ may be chosen.

As the ambient temperature surrounding the cold junction (T_2) varies, a thermally generated voltage appears and produces an error in the output. However, an automatic equal and opposite voltage is introduced in series with the thermal error. This cancels the error and maintains the equivalent reference-junction temperature over a wide ambient temperature range with a high degree of accuracy. When copper leads are integrated with the cold junction, the thermocouple material itself is not connected to the output terminal of the measurement device; secondary errors are thereby eliminated.

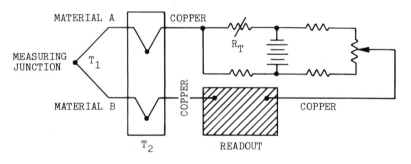

Figure 10–10 Thermocouple Reference Junction *(Courtesy, Omega Engineering, Inc.*

Thermoelectric Refrigeration Method

The thermoelectric ice-point reference chamber relies on the actual equilibrium of ice and water and atmospheric pressure to maintain several reference wells at precisely 0°C. The wells are immersed in a sealed cylindrical chamber containing pure air-saturated water. The chamber outer walls are cooled by thermoelectric cooling elements to cause freezing. An ice shell on the cell wall is sensed by the expansion of a bellows which operates a microswitch, de-energizing the cooling element. The alternate freezing and thawing of the ice shell accurately maintains a 0°C environment around the reference wells.

Completely automatic operation eliminates the need for frequent attention required of common ice baths. Thermocouple readings may be made directly from ice-point reference tables. One can use any combination of thermocouples with this instrument by simply inserting the reference junctions in the reference wells. Calibration of other type temperature sensors at 0°C may be performed as well.

Heated-Oven Method

The double-oven type employs two temperature-controlled ovens to simulate ice-point reference temperatures. Two ovens are used at different temperatures to give the equivalent of a lower reference temperature differing from the temperature of either oven. For example, leads from a chromel-alumel thermocouple probe are connected with a 150° oven to produce a chromel-alumel and an alumel-chromel junction at 150°F (2.66 mV each).

The voltage between the output wires of the first oven will be twice 2.66 mV or 5.32 mV. To compensate for this voltage level, the output leads (chromel and alumel) are connected to copper leads within a second oven maintained at 265.5°F. This is the precise temperature at which chromel-copper and alumel-copper produce a bucking voltage differential of 5.32 mV. Thus, this voltage cancels out the 5.32-mV differential from the first oven, leaving 0 mV at the copper output terminals. This is the voltage equivalent of 32°F (0°C).

The Thermocouple Reference Junction in Application

In figure 10–11 the thermocouple probe is inserted into one end of the junction. A copper extension wire connected to a potentiometer or readout device is plugged into the opposite end of the reference junction.

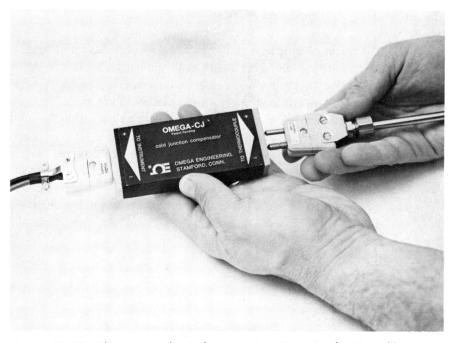

Figure 10–11 Thermocouple Reference Junction Application *(Courtesy, Omega Engineering, Inc.)*

THE RESISTANCE TEMPERATURE DETECTOR (RTD)

You may recall that the thermistor has a negative temperature coefficient. That is, as temperature increases, the resistance of the thermistor decreases. Most conductors of electricity such as the copper wire have a positive temperature coefficient. These conductors increase in resistance as temperature increases. The positive temperature coefficient is termed as *alpha* (α). The thermal component that industry uses is the resistance temperature detector known as the RTD.

RTD Resistance Characteristics

The property of the RTD that is characteristic is its electrical resistance as a function of temperature. This term is known as alpha (α). (Refer to table 10–5 for typical alpha temperature coefficients of RTD materials.) RTDs operate in temperature ranges from $-400°F$ to $+1700°$. The RTD is more efficient than other temperature sensors in that their response to temperature is more linear. A change in temperature will provide an equivalent change in resistance over a

TABLE 10-5. RTD Alpha Coefficients

Alpha Coefficient	RTD Material
0.0038	Copper
0.0039	Platinum
0.0045	Tungsten
0.0067	Nickel

TABLE 10-6. Industrial Worldwide Standards of RTDs

Temperature (F) (in Degrees °)	Resistance (Ohms)
0	93.01
32	100.00
100	114.68
200	135.97
300	156.90
400	177.47
500	197.70
600	217.56
700	237.06
800	256.21
900	274.99

long range of temperatures. The best of the RTDs is the platinum RTD. It has become a world standard in laboratory form for measurement between $-270°C$ and $+660°C$. Precautions and compromises encountered in using other types of electrical temperature sensors are unnecessary. Ordinary copper wire is used to connect the sensor to the readout instrument. Since the calibration is absolute, cold-junction compensation is not necessary. The linear response eliminates corrective networks and errors in interpretation. Freedom from drift makes frequent recalibration unnecessary.

Other standards have been developed which are representative of the resistance values for RTDs at specific temperatures. Table 10-6 lists some of these worldwide industrial standard values.

The RTD Probe (See figure 10–12)

The heart of a typical RTD is the sensing element, made of high-purity platinum wire wound upon a ceramic core. The sensing element is carefully stress-relieved and immobilized against strain or damage.

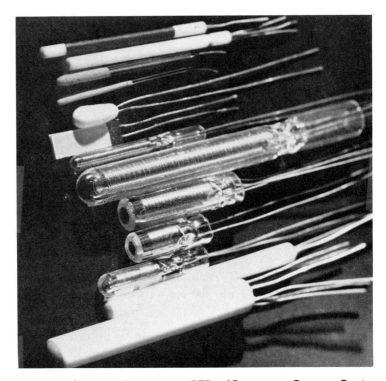

Figure 10–12 Platinum Resistance RTDs *(Courtesy, Omega Engineering, Inc.)*

The standard sensing element is mounted within a stainless-steel sheath, in a manner which provides good thermal transfer and protection against moisture and the process medium. Sheaths are pressure-tight and may often be inserted directly into the process without thermowells. Sheaths are made of stainless steel, silver-soldered or heliarc-welded.

Leads from the thermometer may extend directly from the thermometer for a specified length, may terminate in a connection, or may be connected by a quick disconnect. Figure 10–13 shows these three types of terminations.

The temperature at the lead exit is seldom the process temperature. Standard leads are nickel-plated copper, insulated with extruded Teflon fluorocarbon resin jacketed with wrapped and fused Teflon tape. The seal is high-temperature epoxy. This construction lead will withstand direct submersion in conductive liquids and a maximum temperature of 260°C.

For higher temperatures at the lead exit, leads may be nickel-plated copper insulated with impregnated fiberglass. The seal is ceramic cement or glass

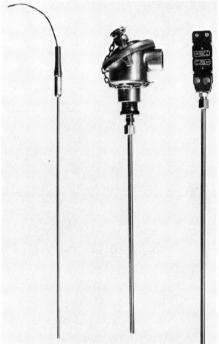

Figure 10-13 Resistance Temperature Detectors (RTD) *(Courtesy, Omega Engineering, Inc.)*

ceramic. This construction will withstand atmospheric moisture and a maximum temperature of 360°C.

RTD Applications (See figure 10-14)

Because of its high electrical output, the RTD furnishes an accurate input to indicators, recorders, controllers, scanners, data loggers, and computers.

Figure 10-14A is a simple form of an RTD bridge used for temperature measurement. The resistance wires of the RTD are used as a part of the sensing branch of the bridge network. The RTD feels the change in temperature while the galvanometer displays the change in relation to fixed resistances. In figure 10-14B one leg of the RTD is connected to the R_3 resistance leg of the bridge. On the lower side of the RTD there are two leads. One lead is tied to the galvanometer and the other to the R_1 resistance leg. The leads on the RTD leg and R_1 resistance leg have equal changes in resistance as temperature changes. Since they are equal changes, there is no effect on the bridge balance. The lead to galvanometer is in the readout path and does not affect the bridge. The hookup in figure 10-14A does have the problem of bridge imbalance.

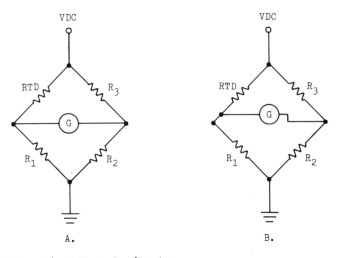

Figure 10–14   The RTD in Application

The major advance in RTD elements is with thin-film operation (see figure 10–15). The element is manufactured in much the same manner as electronic integrated circuits. This precision element, as can be seen in the illustration, is not much larger than the point of a pencil. It is calibrated to standards; most types are available off the shelf. The element may be cemented, taped, or embedded to provide thermal contact.

Figure 10–15   Thinfilm RTD Elements *(Courtesy, Omega Engineering, Inc.)*

Selecting an RTD

Each RTD is made of different material and requires different selection requirements. The selection of a platinum RTD is provided below as a typical selection process.

Nominal resistance—100 ohms at 0°C is standard and stocked for immediate delivery. 200 and 500 ohms are available on special order. Alpha temperature coefficient of resistance change is .00392 ohm per ohm per °C.

Sensitive material—Platinum, reference-grade (99.999 + percent pure) essentially stain-free.

Temperature range—To 750°C.

Interchangeability between units— ± 0.25°C, at 0°C. Closest inter-changeability may be specified at other temperatures.

Long-term stability—Drift less than ± 0.5°C when used to a limit of 500°C.

Sensing-element dimensions—Standard 100-ohm unit, diameter 0.09 inch, length 0.28 inch.

Response time—In ¼-inch-diameter sheath immersed in water flowing at 3 feet per second, response time to 63 percent of a step change in temperature is less than 50 seconds.

SURFACE–TEMPERATURE PROBES

There are a number of unique temperature probes designed to measure the precise temperature of moving and stationary surfaces along with companion readout meters. These meters are called *pyrometers* (high-temperature reading meters).

The probes shown in figure 10–16 are suitable for measurement on flat, concave, and convex surfaces. Surfaces may be metallic or nonmetallic. The

Figure 10–16 Surface Temperature Probes *(Courtesy, Omega Engineering, Inc.)*

probes may be used for molds, plates, walls, glassware, dies, bearings, and other stationary surfaces.

The probes are also used to measure moving and rotating smooth surfaces. The velocity of the surface is a parameter (example: 100 feet per minute). These devices are ideal for measurement of outside temperatures of smooth plastic and steel pipe and tanks.

Some pyrometers are used for measuring layers of plywood, plastic, rubber, paper, and various laminate materials.

Others are used as flat leaf-type temperature probes.

THERMISTORS

The thermistor is a solid-state device that decreases in resistance as temperature increases. The word "thermistor" is derived from two words: *thermal* and *resistor*. In circuit, the decrease in resistance also means an increase in current flow. Thermistors are extremely sensitive. Some may decrease in resistance as much as 5 percent for each degree (centigrade) rise in temperature. Thermistors are made from metallic oxide crystals by means of the *sintering process* (to be explained in ensuing paragraphs).

Some Thermistor Concepts

Thermistors offer extreme sensitivity to temperature differences. For example, a thermistor with practical resistance levels will demonstrate a resistance ratio of 30 between $+25°C$ and $+125°C$, and a ratio of 9 between $0°C$ and $+50°C$. So while thermistors are nonlinear, they offer extreme sensitivity to small temperature changes.

Thermistors offer significant advantages in terms of matching impedance levels to available instrumentation or compensation circuit needs. For example, it is possible to choose a sensor with a $4\%°C$ or greater sensitivity in resistance levels of 100 ohms to 30 megohms at $25°C$. Thermistors offer significant fabrication advantages because they may be mounted in a great variety of substrates. Glass beads and probes are available with ODs as small as 0.014 inch. Discs for measurement purposes are available in small diameters ranging from 0.075 to 0.095 inch.

Conversely, very large measurement assemblies for specialized service can be fabricated with 3-inch diameters and lengths to 6 inches to develop special response characteristics.

Repeatability of measurements in a range of $0.001°C$ is easily achievable with long-term reproducibility of $0.005°C$. Short-term reproducibility is frequently so good that measurement instrument error is the primary source of any uncertainty.

The principal disadvantages of thermistors for measurement applications are significant sensitivity changes (4.9 percent at 0°C, 4.1 percent at +30°C, and 3.8 percent at +50°C), and significant nonlinearity in absolute resistance sensitivity per degree (e.g., at 0°C, 486 ohms per °C; at 30°C, 100 ohms per °C; and at +50°C, 39 ohms per °C).

Thermistor Materials

Raw materials are selected either as chemically pure compounds or as pure compounds that need further preparation before the process is completed. In many cases, the further preparation is for purposes of "mechanical mixing" rather than purity.

The atmosphere in which the sintering goes forward may grossly affect the reaction rate and consequent electrical characteristics. Small quantities of some organic gases may also profoundly affect end performance. The sintered body is polycrystalline and may be porous. Its density is usually lower than a single crystal of the same material. Predicting electrical characteristics from previous identical chemical composition is foolhardy. The homogeneity of the "mix" is not sufficiently predictable; the sintered mass, surface, and deep grains may contain different concentrations of the chosen components.

Metallic oxides such as manganese (Mn), nickel (Ni), cobalt (Co), copper (Cu), iron (Fe), and uranium (U) are used throughout the industry.

Thermistor composition in the specific $NiMn_2O_4$ crystal may be modified by the addition of other material; e.g., Cu, Fe, Co. These additions control the resistivity of the oxide by producing variations in the defect structure of the lattice.

The indirect effect of these additions is the control of the sintered unit by acceleration or retardation of the sinter rate. Other inert additions have the effect of "diluting" the resistivity and/or affecting bonding. Al_2O_3, Kaolin, and bentonite produce an increase in the resistance level, while glass will assist in bonding through its fluxlike action.

The Manufacturing Process (See figure 10–17)

As depicted in the drawing, thermistors are manufactured from special formulations of powdered metal oxides compressed into a small disc and then sintered. The mix of metals, sintering temperature, and atmosphere determine the thermistor slope and characteristics. The sintered disc is coated with silver

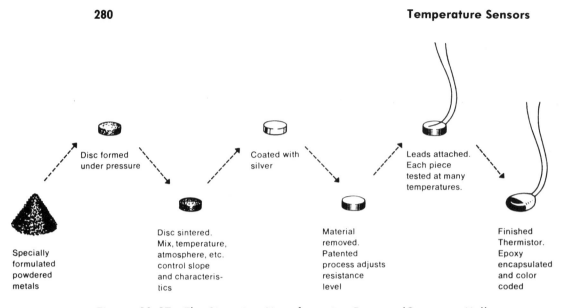

Figure 10-17 The Sintering Manufacturing Process *(Courtesy, Yellow Springs Instrument Co.)*

and a patented process of material removal adjusts the resistance level so that all thermistors of a given type and value display nearly identical characteristics. Leads are attached and each thermistor is individually tested at several different temperatures. The finished thermistor is epoxy-encapsulated for ruggedness and then color-coded for resistance-value identification.

The sintering process is the most critical process of thermistor manufacturing.

The process may be defined as the densification and bonding of molded parts at temperatures well below the melting point of the individual components. In addition to the mechanical processing, thermistors also depend on sintering to promote and control the solid-state reaction of a mixture of components. Cooling rate after the sinter may contribute to phase stability, aging characteristics, and chemical stability.

Thermistor in a Bridge Application (See figure 10-18)

Bridge design, while susceptible to more rigorous analysis and calculation, can be generalized for less stringent requirements, as we will see.

Resistance of the detector should be greater than the thermistor resistance at the lower temperature of the bridge design. A very simple computation of a nearly optimized bridge follows.

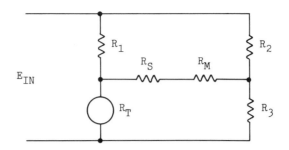

Figure 10–18 Thermistor in a Bridge Application

Choice of the sensor impedance depends largely on the temperature to be measured. For obvious reasons, an attempt should be made to choose a sensor with reasonable resistance levels at the temperature of interest.

The bridge is at null when $R_T = R_1$.

For maximum linearity, $R_2 = R_3 = R_T$ at mid-point of temperature range.

Bridge voltage should be chosen considering self-heat/dissipation. The maximum voltage across the thermistor (E_{Th}) may be determined by $E_{in} = PR_T$ where P is the dissipation allowable for the acceptable self-heat error and R_T is the thermistor resistance at mid-range.

Concerning meter selection circuitry ($R_m + R_s$), if ammeter readout is desired, $R_m + R_s$ should be approximately 10 times the R_{TH} at the maximum temperature (lowest thermistor resistance). Use a constant current source instead of constant voltage or current bridges.

For voltmeter readout, $R_m + R_s$ should be at least 10 times the R_{TH} at the lowest temperature (highest thermistor resistance).

Thermistor Applications (See figure 10–19)　In figure 10–19A, the thermistor is used as a simple temperature-measuring device. Here the thermistor is placed in series with a meter. The meter is current sensitive and calibrated so that a change in current will cause an equivalent change in degrees of temperature.

In figure 10–19B, the thermistor is placed in a bridge network to provide a more precise measurement. A meter (galvanometer) is placed across the bridge to monitor minor changes in current which are reflected as degrees of temperature on the meter.

In figure 10–19C, a pair of thermistors are placed in two physical locations while being tied electrically to a bridge network to compare temperatures. The bridge is then balanced to a no-current-flow condition. From the balanced condition, any future change in temperature will cause a change in current flow through the meter (galvanometer). Since the meter is calibrated in degrees, the

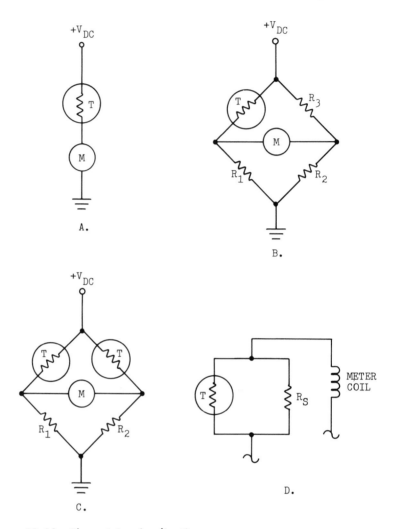

Figure 10-19 Thermistor Applications

amount of temperature change will cause the calibrated meter to reflect the change in terms of temperature.

In figure 10-19D, a thermistor is placed in parallel with a shunt resistor and in series with a meter coil. The thermistor has a negative temperature coefficient. The copper wire has a positive temperature coefficient. The thermistor and shunt are chosen to have equal and opposite coefficients so that a change in temperature will affect both but in opposite directions. This action provides the effect of temperature compensation.

Selecting the Correct Thermistors   Selection of a suitable thermistor resistance value for a given application is generally based on the following considerations:

Temperature span—Resistance values of 100K to 500K Ω@ (25°C) are used for high temperatures (300°F to 600°F). Resistance values of 2K to 75K Ω@ 25°C are usually used for temperatures between 150°F and 300°F. Thermistors having resistance values of 2K to 5K Ω@ 25°C are recommended for temperatures between 32°F and 212°F. Low resistance values (100 to 1,000 Ω@ 25°C) are used at low temperatures in the range minus 100°F to plus 150°F.

Resistance values at temperature-span extremes—These may be taken from resistance-temperature tables. Factors to be considered are the following:

1. Maximum resistance at low temperature must not be too high to meet the needs of associated circuitry such as amplifier, readout, etc. If resistance at low temperature is very high, the possibility of spurious signal pickup must be considered. If high resistance is required for other reasons and pickup is a problem, then the use of shielded lines, filters, or dc power may be desirable.
2. Minimum resistance at high temperature must not be too low to meet the needs of amplifier, readout, etc. If resistance at high temperature is too low, consideration must be given to possible errors caused by contact resistance, line resistance, and line-resistance variation with changes in ambient temperature.

Sensitivity—Percent change in resistance per degree F or C over the selected temperature span is an important consideration.

Self-heat—Power (I^2R) dissipated in the thermistor will heat it above its ambient. The temperature increase is a direct function of the dissipation constant of the thermistor in its mounting, in its operating ambient environment. Under fixed conditions, an offset allowance may be made for self-heating, and it need not be allowed for in the tolerance requirements. Note, however, that anything that affects the dissipation constant (such as velocity of the medium being measured, etc.) will change this offset.

Specifying tolerance—Most applications have a tolerance expressed in temperature units. On the other hand, thermistors are usually specified in terms of resistance tolerance. It is characteristic of thermistors that a fixed resistance tolerance over a temperature span is equivalent to a tapering temperature tolerance that is smaller on the low temperature end and larger on the high temperature end. The thermistor comes in many forms, as is illustrated in figure 10-20.

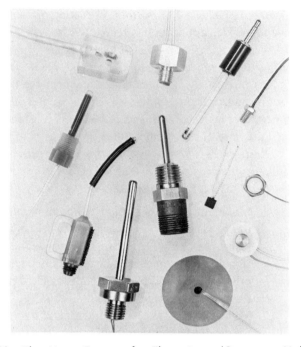

Figure 10-20 The Many Forms of a Thermistor *(Courtesy, Yellow Springs Instrument Co.)*

REFERENCES

Reference data and illustrations were supplied by state-of-the-art manufacturers.

Permission to reprint was given by the following companies:

Thermocouples and resistance temperature detectors (RTDs)—Omega Engineering Co., Inc., Stamford, Connecticut

Thermistors—Yellow Springs Instrument Co., Inc., Yellow Springs, Ohio

All copyrights © are reserved.

11

Meteorological Sensors

No book on transducers would be complete without some information about meteorological sensors. Although this book is generally directed at industry, we shall take a cursory look at those sensors that are used to monitor and predict weather.

THE WIND-SPEED SENSOR

The photograph in figure 11-1 is a Texas Electronics, Inc. Model TV-110-L2 wind-speed sensor (anemometer). The version you see is a three-cup version, although six-cup versions are common. The six-cup version is particularly suited for applications involving low-wind-speed studies since its starting threshold is lowered by the three additional cups.

Exposed components are constructed of gold anodized aluminum to retard corrosion and to maintain a rugged, lightweight, sensitive assembly.

The anemometer shaft is directly connected to a sixty-slot disc which, when rotated, interrupts the infrared beam produced by a light-emitting diode. The interrupted signal is detected by a photosensitive transistor located on the opposite side of the disc. Sealed low-torque ball bearings are used to make periodic inspections unnecessary. Since there is no frictional coupling to the transducer, drag is kept at a minimum.

The output signal of the sensor is a pulse train whose repetition rate is linearly proportional to wind speed. A signal conditioner/power supply mounted remotely from the sensor shapes, amplifies, and converts the low-level pulsed sensor signal to an analog dc voltage that is proportional to wind speed.

The wind-speed sensor is often used with a control system. A wind-speed activated control system is designed for switching on and/or off, various types

Figure 11-1   The Wind-Speed Sensor *(Courtesy, Texas Electronics, Inc.)*

of equipment such as deodorizers, air samplers, alarms, beacons, etc. when wind speed exceeds or falls below a certain selectable control value. The control output function (either switch-on or switch-off, or both) can be set for activation at any desired wind-speed value from 0 to 100 mph. An adjustable time delay for each set point can be made available which minimizes premature on–off cycling of the controlled equipment when wind speed is fluctuating above or below the selected preset value.

Two modes of operation are provided—automatic and latching. In the automatic mode, the relay (s) is energized at wind speeds exceeding the set point (s), but automatically de-energizes when wind speed falls below the set point (s). In the latching mode, the relay (s) energizes when the wind-speed set point (s) is exceeded and remains energized until it is manually reset.

WIND-DIRECTION SENSOR

The Texas Electronics, Inc. Model TD104P wind-direction sensor provides wind-direction presentation from 0° to 360° (see figure 11–2). The sensor utilizes a long-life plastic potentiometer that is mechanically connected to the wind vane shaft. The dc resistance obtained at the potentiometer wiper contact is directly proportional to the angular position of the wind vane. The wind vane is exceptionally long (33 3/4 inches). This length is utilized to minimize errors caused by overshoot and inertial action. All exposed components are constructed of lightweight gold anodized aluminum to resist the elements. The rotating unit is permanently sealed. Lubricated ball bearings support the vane and vane shaft.

A remotely mounted signal conditioner and power supply provides a dc signal which is proportional to an azimuth range of 0° to 360°.

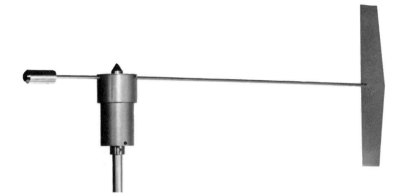

Figure 11-2   The Wind-Direction Sensor *(Courtesy, Texas Electronics, Inc.)*

The wind-direction sensor is often used with a control system. A wind-direction-activated control and indicating system is designed to energize or de-energize process equipment for controlling a pollution nuisance whenever the wind shifts to a critical direction sector. Deodorizers, high-volume air samplers, incinerators, dust fallout regulators, etc. can be activated or deactivated automatically as desired. The control output function (either switch-on or switch-off) can be set for activation at any desired wind-direction zone (zone degrees or more in width) over a range of 0° to 360°. An adjustable time delay incorporated into the circuit minimizes premature on–off cycling of the controlled equipment when the wind vane is fluctuating in and out of either edge of the direction sector. Another adjustable time-delay unit holds the circuit "on" once it has been activated so that it will remain in the "on" position for a selectable number of seconds. Delay durations are adjustable by two knobs located on the console panel.

When used with a wind-speed controller, the two systems can be series-interconnected to give a variety of control options of high or low wind speeds in or out of a chosen sector.

THE RELATIVE HUMIDITY SENSOR

Figure 11-3 is a photograph of the Texas Electronics, Inc. Model TH-2013 humidity sensor and Model 210 signal conditioner and power supply.

The sensor assembly contains a hydroscopic inorganic sensing element. Its expansion and contraction positions the suspended core of a linear variable

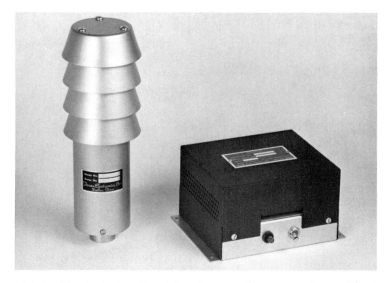

Figure 11-3   The Relative Humidity Sensor *(Courtesy, Texas Electronics, Inc.)*

differential transformer (LVDT). The absence of friction-inducing linkages and wiping contacts minimizes hysteresis and improves accuracy. The LVDT output signal, when processed, is directly proportional to relative humidity. Four nature-shaded pagoda fin cups effectively surround the sensing element and permit free flow of air about the sensor, while protecting it from rainfall. The LVDT and associated components are enclosed by a tubular skirt located below the sensing element. Nonferrous materials are used in the construction of the sensor housing to guard against rust and corrosion. All aluminum parts are gold-anodized.

Sixty-hertz input power to the signal conditioner is stepped down, rectified, and regulated so that the final indicated signal is unaffected by fluctuations in line voltage. The regulated power energizes a low-voltage ac oscillator whose output is fed to the sensor assembly LVDT. The returning signal from the LVDT is detected, filtered, and balanced in the signal conditioner and is then presented as a processed signal to the output terminals. A two-conductor cable carries the signal to applicable recorder, data logger, indicator, etc.

THE BAROMETRIC PRESSURE SENSOR

Illustrated in figure 11-4 is the Texas Electronics, Inc. Model TB-2012 barometric pressure sensor and its Model 2012 signal conditioner and power supply.

Figure 11-4 The Barometric Pressure Sensor *(Courtesy, Texas Electronics, Inc.)*

The sensor is made up of a bellows which is directly coupled to the core of a linear variable differential transfomer (LVDT). No physical contact is made between the core and transformer, thus eliminating friction. The absence of friction-inducing linkages or potentiometric devices improves resolution and minimizes hysteresis. Also enclosed in the sensor housing is a temperature-sensitive resistive element for temperature compensation. The LVDT is excited from the signal conditioner module.

Sixty-hertz input to the signal conditioner is stepped down, rectified, and regulated so that the final indicated signal is unaffected by fluctuations in line voltage. The regulated dc power drives a low-voltage oscillator whose output is fed to the sensor-transmitter module. The signal returning from the sensor-transmitter module is detected, filtered, and balanced in the signal conditioner module, and the processed signal is presented to the output terminals.

AIR-TEMPERATURE SENSOR (See figure 11-5)

The Texas Electronics Model TT-101 air-temperature sensor consists of a highly sensitive linear thermistor-resistor network within a protective housing.

Air temperature variations create a resistance bridge imbalance; the

Figure 11–5   Air-Temperature Sensor
(*Courtesy, Texas Electronics, Inc.*)

subsequent output signal varies linearly with temperature. A naturally
aspirated sensor shelter is provided which permits temperature measurement
substantially free of solar radiation. Exposed shelter components are constructed of anodized aluminum for maximum environmental protection. The
signal conditioner output voltage may be interfaced with various types of
recorders, indicators, data loggers, etc. as required by the user.

Two or more sensors may be mounted on a tower to obtain vertical
temperature profile studies of the measurement of inversion conditions. This
differential air temperature (ΔT) may be displayed or programmed in the same
formats as a single sensor.

Power requirements are 115 Vac, 60 hertz.

RAINFALL TRANSMITTER (See figure 11–6)

There are many types of rainfall transmitters. The most desirable are those with
accuracy during heavy rainfall. The Texas Electronics, Inc. Model 6118–1 is a
state-of-the-art device for this function.

The transmitter basically consists of a collector and series of funnels to
divert the rainwater to a tipping-bucket mechanism. The collector and tipping
bucket are so designed that each hundredth of an inch of rain causes the alternate fill and tip of the mechanism. A sealed glass enclosed mercury switch is attached to the mechanism so that an electrical impulse is generated with each tip
of the bucket. This signal is then transmitted via a two-conductor cable to a
receiving mechanism. The receiver can be a counting mechanism, an event
recorder, telemetering equipment, data-acquisition system, etc. The transmitter
is fully automatic and requires no attention. Each bucket, as it dumps,

Figure 11–6 Rainfall Transmitter
(Courtesy, Texas Electronics, Inc.)

discharges its water through the bottom of the unit, eliminating periodic hand dumping or measuring.

A 120-vac power supply or a battery (12 vdc) can be used to apply the power to produce the required pulse. Battery power plus a spring-wound recorder provide an ideal solution to remote operation where central station power is not available.

Adjustable leveling pads are provided so that the transmitter may be installed in an absolutely level position. Calibration accuracy is adversely affected, in any tipping-bucket mechanism, by an out-of-level installation.

The component parts of this transmitter are constructed primarily of aluminum and brass for long trouble-free operation. All aluminum components are gold-anodized except the outer housing, which is primed and painted white.

The receiver orifice is a precision complex spun assembly with a knife edge, resulting in an extremely accurate sensor. Screens are provided on all openings to keep leaves, bugs, etc. from entering the mechanism. The tipping bucket itself is constructed of brass.

THE METEOROLOGICAL SENSOR ARRAY (See figure 11–7)

Probably the most sensible way to purchase meteorological sensors is in an array. Arrays can be made to handle any combinations of measurable data such as the following:

1. Horizontal wind direction
2. Wind speed
3. Barometric pressure
4. Relative humidity
5. Ambient temperature
6. Vertical temperature differential
7. Precipitation

The arrays are usually purchased along with a rack-mounted power supply and signal conditioner for remote data collection.

The Texas Electronics Series 3000 is typical of these high-technology meteorological systems. This model consists of all the sensors listed above along with a modular plug-in signal conditioner and power supply. The plug-in module concept allows choice of a system package that fits a particular application.

Power for the array and signal conditioner is 120 Vac, 60 hertz, or 12 Vdc.

Electrical analog signal outputs from each sensor / signal conditioner combination are converted to linear dc voltages proportional to the parameter being

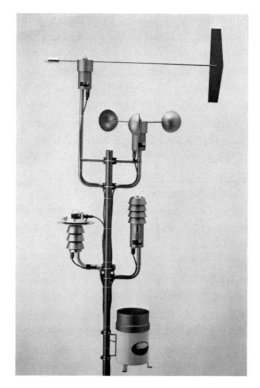

Figure 11–7 Meteorological Sensor Array *(Courtesy, Texas Electronics, Inc.)*

monitored. The variable voltage signals are scaled to read direct in millivolts dc or ranged 0 to 1.0 Vdc as required. Options include dual signal outputs for each parameter.

REFERENCES

Reference data and illustrations were supplied by a state-of-the-art manufacturer.

Permission to reprint was given by the following company:

Meteorological sensors—Texas Electronics, Inc., Dallas, Texas
All copyrights © are reserved.

Index

ABINGDON COLLEGE OF FURTHER EDUCATION

Library